Eugenija Baksiene

Lakes sapropel for soil fertilization

Eugenija Baksiene

Lakes sapropel for soil fertilization

LAP LAMBERT Academic Publishing

Impressum / Imprint

Bibliografische Information der Deutschen Nationalbibliothek: Die Deutsche Nationalbibliothek verzeichnet diese Publikation in der Deutschen Nationalbibliografie; detaillierte bibliografische Daten sind im Internet über http://dnb.d-nb.de abrufbar.

Alle in diesem Buch genannten Marken und Produktnamen unterliegen warenzeichen-, marken- oder patentrechtlichem Schutz bzw. sind Warenzeichen oder eingetragene Warenzeichen der jeweiligen Inhaber. Die Wiedergabe von Marken, Produktnamen, Gebrauchsnamen, Handelsnamen, Warenbezeichnungen u.s.w. in diesem Werk berechtigt auch ohne besondere Kennzeichnung nicht zu der Annahme, dass solche Namen im Sinne der Warenzeichen- und Markenschutzgesetzgebung als frei zu betrachten wären und daher von jedermann benutzt werden dürften.

Bibliographic information published by the Deutsche Nationalbibliothek: The Deutsche Nationalbibliothek lists this publication in the Deutsche Nationalbibliografie; detailed bibliographic data are available in the Internet at http://dnb.d-nb.de.

Any brand names and product names mentioned in this book are subject to trademark, brand or patent protection and are trademarks or registered trademarks of their respective holders. The use of brand names, product names, common names, trade names, product descriptions etc. even without a particular marking in this work is in no way to be construed to mean that such names may be regarded as unrestricted in respect of trademark and brand protection legislation and could thus be used by anyone.

Coverbild / Cover image: www.ingimage.com

Verlag / Publisher:
LAP LAMBERT Academic Publishing
ist ein Imprint der / is a trademark of
OmniScriptum GmbH & Co. KG
Heinrich-Böcking-Str. 6-8, 66121 Saarbrücken, Deutschland / Germany
Email: info@lap-publishing.com

Herstellung: siehe letzte Seite /
Printed at: see last page
ISBN: 978-3-659-77126-2

LAKES SAPROPEL FOR SOIL FERTILIZATION

Eugenija Bakšienė

2015

Content

Introduction

It is very important to preserve organic soil material of light textured soils. Even if extensive agriculture is being carried on, the preservation of nutritious possible only with its gradual replenishment with organic fertilizers. With agriculture becoming more intensive and increasing production, humus is rapidly mineralised and the soil becomes exhausted. It can be amended applying various organic fertilizers. In the world of all organic waste returned into soil up to 70 % are plant waste, 23 % industrial waste, and only 1 % organic substances of other origin. Large amount of organic matter accumulated in lakes can be included into this one percent. It abundantly accumulates in lakes situated in the regions with unproductive soils. The life of human beings caused water resources to be destroyed because of negative human life itself. Therefore natural eutrophication lakes and anthropological activity of people in an area of lakes, in the world many lakes are silty, they are decaying and turning into marsh.

Lithuania is know as the"land of lakes" and they number 2850 (>0,5 ha in area). As lakes are silting the concentration of phytoplankton increases. Such lakes are suitable neither for recreation nor for pisciculture; they can even become the source of infection diseases. There is plenty of means to reduce lake silting. Organic sediments accumulated in old lakes is a fossil that has not been usedwidwly yet. Its layers are usually 6-8 m deep, sometimes even 24 m deep. The deposits of sapropel in lakes are about 5,8 miliard m^3. Nearly the same quantity in former lakes – marchlandsof lacustrine origine.

Dealing with the already silted lakes most attention is given to hidrotechnical and mechanical cleaning – removal of lake sediments. If they are properly used come costs involved in the cleaning could return. Application of lake sediments as fertilizers looks most apporopriate. Besides, the highest numbers of silted lake are in the zone with unfertile, fine textured soils where requirement for fertilization is the highest. In lake sediments (sapropel) the amount of bacteria decomposing cellulose is not high; therefore, the functioning of lake sediments is more prologed than of other organic fertilizers. It is of utmost importance in cultivated soils.

Fertilization with lakes sapropel enriches the soil with organic matter as well as improves its atructure, physical properties; it is also an inportant measure preventing wind erosion of soil. Practically all investigatiors agree that lake sediments function as long-term measure for improving agrochemical and physical properties of soil. Research carried out in various countries on the effects of lake sapropel suggests that its efficacy depends on the chemical composition. The effect of soil sediments highly increases if they are composted with manure. The amount of microorganisms in such sediments raises and consequently does the amount of nutritious substances.

Information in the book updates knowledge is relevant to increasing the understanding of application of lakes sapropel for soil fertilization and in the fulfilment of general tasks of agronomy.

1. Lakes of Lithuania

In Lithuania there are 2 850 lakes larger than 0.5 ha and 3 150 lakes of natural origin smaller than 0.5 ha. There are also 3 000 artificial lakes. Altogether the lakes occupy 920 km² of Lithuanian territory. The average number of lakes in the area unit is 1.4 %. If this parameter is calculated by the data of only 932 lakes larger than 10 ha, it would be 1.3 %. Most abundant in lakes are Baltic highlands that extend from Vištytis to Drūkšiai lake, which is the largest in Lithuania (4500 ha). The highest numbers of lakes are located in Aukštaičiai Upland, in northeastern lake district. Here smaller lake districts – Zarasai, Dubingiai, Molėtai, Ignalina-Kaltanėnai - are located. In Dzūkija and Sūduva highlands a relatively small Trakai, Daugai, Veisiejai lake districts are located. The second area according to the lake abundance is Žemaičiai Highlands, the third one is Southeast plains with small lakes located there. Much less lakes are located in Central and Seaside Lowlands (Kilkus, 1991, 1993, 1998).

Fig. 1. National Parks of Aukstaitija and Trakai.

After the last glaciation on the territory of Lithuania there were many more and larger lakes than it is now, and their decline and disappearance continues. There are no young lakes in Lithuania. All of them are mature because everywhere the sandbank is already formed (Bieliukas, 1956, 1961).

Lake is a natural inland water body formed in surface waterholes. Waterhole form of the largest Lithuanian lakes is usually close to the cone shape, the form of small lakes is hemisphere.

The water level in lakes is constantly changing. Short-term water level fluctuations are related to the weather conditions and occur in all lakes. Seasonal water level changes are associated with the change of seasons and are typical of inflow and outflow lakes. The largest many-year water level fluctuations are determined in Sartai and Šventas lakes. Here the recorded water level fluctuation reached up to 1.5-2.2 m. In Lithuania, like in other lakes of middle latitudes, the highest water level occurs in spring after the thaw and flood in rivers as well as in autumn, when precipitation is most abundant. Many-year fluctuations of water level are associated with long-term climate changes. This kind of fluctuation in particularly characteristic to stagnant lakes. Long-lasting water level fluctuations are related to geological factors, rise of lake bottom, sediment accumulation (Garunkštis, 1975).

From a distance, water mass of Lithuanian lakes looks blue, but the water is mostly of yellow-green colour. The colour depends on the amount of humic acids getting into water and on the accumulated organism species (Bieliukas, 1961; Garunkštis, 1988).

In Lithuania, lakes of medium mineralization (200-500 mg l^{-1}) with hydrocarbonate calcium water dominate. Mineralization is constantly changing because of changing leakage of drainage basin, chemical and physical properties of water, occurring sedimentation processes. Differences between water mineralization levels in spring and winter are the largest. In spring, snow melt water of low mineralization gets into lakes, and in winter lakes are supplied with more mineralized underground water (Kilkus, 1987). In sandy Southeast plain and Žemaičiai upland lakes of low mineralization (less than 200 mg l^{-1}) prevail. There are also some lakes with water of particularly low mineralization (only about 50 mg l^{-1}). Usually, they are surrounded by bogs or formed in areas with very infertile soils. Lakes of karst region in northern Lithuania are most mineralized (500-1000 mg l^{-1}), rich in sulfates and chlorides (Garukstis, 1975).

The water-heat regime depends on water mineralization. The higher is the mineralization, the lower is the heat intensity. But heat distribution in lakes is mostly predetermined by meteorological, hydrological, hydraulic and morphological factors and direct solar radiation. The heat exchange through the water surface, amount of heat getting into the lake and its redistribution according to the depth depend on them. Besides the direct sunlight, the inflow and outflow lakes receive a certain amount of heat with river waters. A lot of heat is released from decomposition of organic material in the bottom of the lake (Kilkus, 1991).

In summer, warm water layer forms at the surface, and cold water layer – at the demersal zone of lakes. When the water is fresh and at the maximum temperature of water density, i.e. about + 4 °C, directly stratified vertical thermal structure forms. Uneven horizontal distribution of water temperatures could be caused by wind, water from tributaries and human economic activity in lakes and ponds used for energy purposes.

There are periods in spring and autumn when water is mixed and water temperature becomes equal through the whole lake. Water temperature of all layers in majority of Lithuanian large and shallow lakes is approximately equal.

Lake water contains gases: oxygen, nitrogen, carbon dioxide, argon and others. Gas accumulation in water does not depend on mineralization, but more on the lake depth. Gases from the atmosphere into water or from water to the atmosphere get directly in the upper layers of the lake. Water temperature and atmospheric pressure are very important for gas entry and distribution. Into the deeper layers, gas enters through diffusion: first into middle water layers and then into the deeper layers. In addition, with currents the surface water with the absorbed gas sinks deeper (Bieliukas, 1961).

All bodies abound in great variety of organisms. According to the arrangement of organisms, richness in nutrients, their biocoenoses and transformation, lakes are divided into several groups: 1) Oligotrophic – very deep lakes, with clear and cold water. They are nutrient-poor. Such lakes are little silted and are most widespread in

Scandinavia; 2) Mesotrophic – the deepest Lithuanian lakes of moraine, hilly landscape (Dusia, Galvė, Plateliai, etc.). They are characterized by slightly silty shores and higher diversity of flora and fauna; 3) Eutrophic lakes make strong contrast to oligotrophic ones. These are shallow lakes, warming up to the very bottom, nutrient-rich and abounding in flora. In such lakes intensive decomposition of organic matter and sedimentary processes take place. The water in such lakes is often oxygen-poor (Žuvintas, Amalvas, Ofelija, etc.); 4) Dystrophic – small lakes in peat bogs with undergoing intense decomposition of organic residues, associated with peat formation. Only most resistant fish (roach, pike, perch, carp) could live in such lakes (Bieliukas, 1961).

Every Lithuanian lake is unique. Lakes differ in their origin, age, water characteristics, landscape beauty; different plants and fish inhabit them. An important parameter characterizing the recreational potential of Lithuania is the total length of the coastline. It reaches 6 240 km, i.e. about 2 km of lakesides for 1 thousand of the inhabitants (Kilkus, 1987). However, part of the lake sides are silty and not suitable for recreation.

2. Sedimentation processes in lakes

The formation lake sediments – sapropel – is influenced by a number of factors: geochemical composition of the area, biochemical physiology of water, forms of water movement, bacterial and chemical processes as well as diversity of lake flora and fauna. Separate sedimentation factors affect the processes only at a certain depth zone or in different zones with varying degree. Primary sedimentary materials are divided into allochthonous and autochthonous, what roughly correspond to minerologic and organic materials. Minerologic material occurs mainly by inflow and then due to eroding lake water activities, underground springs and, ultimately, the direct process of sedimentation in a basin. Primary organic sediments are mainly formed in the lake itself. It is living creatures at the bottom, fish and plankton

(Belenkij, 1981; Lopotko and Jevdokimova, 1986). Not all organic material of planktonic origin settles on the bottom of the lake. A substantial part of it is decomposed under the influence of microbiological, hydrochemical, hydrophysical and hydrodynamic processes. Fragmentation of organic materials primarily depends on the lake, depth of its separate parts. Most organic matter of planktonic origin falls to the bottom in the shallow places of the lake, least – in deep places (Garunkštis and Stanaitis, 1978).

The upper layers of lake sediment, very watery, diluted, not sufficiently formed are called pelogen. In these layers, resulting from the decomposition of dead plankton, the main sapropel formation processes occur (Garunkštis, 1975).

Fig. 2. Lake are silting

Lower lake sediment layers were generally formed in the first post-glacial period, when water in lakes was ice cold and living organisms were very sparse. Therefore, those layers are rich in minerals. Calcareous sapropel is dating those times. It consists of $CaCO_3$, trapped here in the form of a solution from the shores and rivers. With warming climate and emergence of living organisms, sapropels containing more abundant organic matter are formed (Garunkštis, 1975; Kabailiene, 2006).

9

Studies of sedimentation processes indicate that sediment layer depends on the lake type; for example, bottom sediment layer in karst lakes (Lake Kirkilai, Biržai distr.) is very thin, although by sedimentation intensity it should be considerably thicker. This could be caused by close connection of the surface and underground water leading to sediment penetration into underground cavities (Taminskas, 2000).

Depending on the lake nutrition mode and intensity, different load of suspended and dissolved substances, that affect the intensity of sedimentation, gets into the lake. Significantly enhanced shallowing is noticeable in shallow lakes, abundantly fed by surface water. During the warm season, the thinner is the water layer, the greater is the opportunity for water to heat up deeper; therefore forming better conditions for phytoplankton, various algae and water vegetation to develop (Ciūnys and Katkevičius, 2008). The deeper is the lake, the less phytoplankton reaches the bottom because it disintegrates going down the water column (Garunkštis and Stanaitis, 1978). Human economic activity also affects the siltation of lakes. Municipal waste waters are particularly damaging to lakes as the lake is contaminated not only with organic material, but also with harmful heavy metals and their compounds.

Surroundings of the lake condition the sediment accumulation. Rivers, rain, snow melt and ground water feed the lake with various dissolved substances. Solid particles – products of mechanical disruption, fall into lakes if soil erosion or abrasion of shores takes place. Residues of deciduous trees growing around the lake also supply the lake with organic matter. However, sedimentation processes in lakes situated in forests are stable. Vegetation reduces surface erosion and inflow of deposits into lakes. Such lakes are usually fed by hydrogen carbonated ground water (Kabailienė, 2006).

In separate lakes the sediment layer can reach up to 31 m (Katkevicius et al., 1998). The speed of their accumulation depends on the amount of vegetable and animal organisms. Over a year, the amount of deposited silt in lakes of various nutrition mode can range from a few to a dozen centimetres. Its composition is not uniform. In some lakes more mineral deposits settle, in others, especially multi-nutritious or fed by wetlands water, more organic and transitional type sediments accumulate. These deposits are initially very loose, literally hanging in the water, but then the processes of their clumping and decay start (Kavaliauskienė, 1997).

Water level fluctuation, seasonal weather changes, differences in the amounts of plankton and terrigenous material getting into lake affect the horizontal stratification of sapropel. During the warm season, when living organisms in the lake are active, sapropel is abundant in organic matter, while in winter and especially spring, the forming thin layers are of more uniform composition and richer in mineral particles (Garunkštis, 1975; Linčius, 1977; Kabailiene, 2001).

3. The composition and properties of sapropel

Sapropel (organic-rich lake mud) (from the Greek language "sapros" – decay, "pelos" – silt) is unconsolidated colloidal gel-like deposits, present in freshwater bodies. It is formed of plant and animal remains which, along with mineral matter, mostly clay and sand as well as oxides of various elements present in water, settle at the bottom of the basin during episodes of reduced oxygen availability, due to the effect of partial oxidation processes (Linčius, 1977; Katkevičius et al., 1998; Lopotko and Jevdokimova, 1986).

All lakes sapropel are subdivided into organic (50-90 % of organic mater), calcareous (30-60% of calcium carbonate), siliceous (25-45% of silicon dioxide) and mixed groups (Fig. 3) (Rubinstein, 1984; Sapropels of BSSR..., 1986). All kinds of lake sapropel are used to fertilize infertile soils.

11

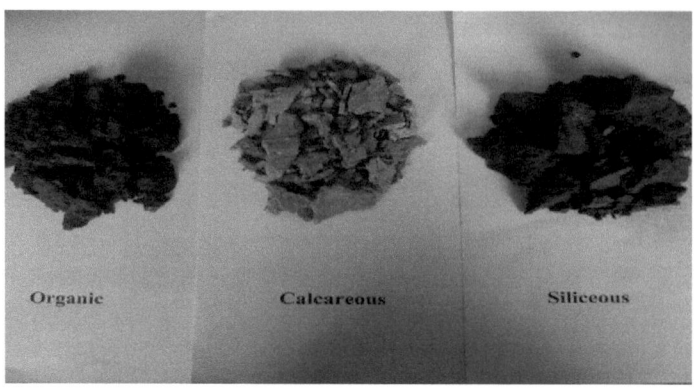

Fig. 3. Various kind of lakes sapropel according different chemical composition

Organic material is the most valuable part of sapropel. Its quantity may vary from 10 to 15 %. The organic material contains 11-38 % of humic acids, 20-31 % of carbohydrates also large quantities of biochemically active substances – amino acids, enzymes, vitamins (A, B_1, B_2, B_{12}, D).

Organic material of sapropel is a nutrient for microorganisms. Therefore, there is a direct connection between microorganisms and the amount of organic matter (Kanopkaite-Rozgiene et al., 1962).

Sapropels are rich in various microflora, especially bacteria. 1 g of dry sapropel could harbour from a few hundred to tens of millions of protein-degrading bacteria, tens of millions of mineral nitrogen assimilating bacteria and ologonytrophyls. Sparse or completely absent are actinomycetes, cellulose degrading microorganisms (Milto et al., 1981). In lake sapropel, microflora is much richer that in sapropel located under the drained peat. In sapropel ventilated in sedimentation tanks, amount of microorganisms increases by 10 times or more (Shinkariova et al., 1981; Grantina-Ievina et al., 2014).

Once in the soil, sapropel induces the activity of microorganisms and consequently the biochemical processes (Milto et al., 1981). Ability of sapropel to retain moisture, change soil acidity strongly influences the development of micromycetes. Because of intense microbial activity, in sapropel-fertilized soil vegetable waste destruction intensifies thus creating better conditions for the plant growth (Lugauskas et al., 1999).

It has been found out that the organic matter complex of sapropel included protein, carbonhydrogens, humus acids, carotins, vitamins of groups B, E, D and other. Many amino acids are found in sapropel: lizine, asparagine, glutamic acid, glycol, histidine, threonine, alanine, tirozine, valine, methionine, phenylalanine, leucine, isoleucine. Sapropel is consided to be valuable for agriculture it has carotene, vitamins D, B_2, B_3, B_{12} and other and also folic acid. It can be judged about the amount of these matters in the sapropel of various chemical composition by data presented in the table 1.

Table 1. The amount of B group vitamins in sapropel (according Katkevicius et al., 1998)

Vitamins	Tipe of lakes sapropel			
mg kg^{-1} of sapropel	Calcareous	Organic	Siliceous	Mixed
Inositol	106,9	86,00	58,12	74,46
Biotin (H)	0,063	0,059	0,065	0,089
Panthothenic acid (B_3)	6,69	15,87	11,11	15,06
Tiamine (B_1)	0,030	0,073	0,017	0,056
Pyridoxine (B_6)	0,091	0,064	0,011	0,363
Nicotinic acid (PP)	3,220	0,647	2,62	3,93
Paraaminobenzoic acid	0,005	0,049	0,010	0,068
Vitamin B_{12}	0,367	0,073	0,539	0,027

It can be seen from data presented on the table 1, that the calcareous saproepl is more rich in vitamins than the siliceous, though the ash-content in the first mentioned sapropel is considerably greater than in the second. There is especially much inositol, which stimulates the growth of the plants in the calcareous sapropel. There are much panthothenic acid, thiamine, nicotinic acid, and especially vitamin B_{12} in the all sapropel of chemical composition.

Other part of sapropel – ash content. The most valuable is sapropel having ash content less than 10 %. Ash quantity and composition depend on conditions of sapropel formation and ranges from 10 to 90 %. The main components of sapropel are nitrogen, phosphorus, potassium, calcium and magnesium. Its chemical composition is often not worse than the manure (Table 2).

There are also: silicon, iron, aluminum, manganese, sodium oxides. Sapropel contains trace elements in various quantities: manganese – 30-100 mg kg^{-1}, zinc – 80-400 mg kg^{-1}, copper – 10-60 mg kg^{-1}, molybdenum – 5-20 mg kg^{-1}, cobalt up to 15 mg kg^{-1} of dry sapropel (Katkevicius et al., 1998).

Table 2. Chemical composition of lakes sapropel and manure

Substances	% dry matter				
	N	P	K	Ca	Mg
Calcareous lake sapropel	0,62	0,02	0,03	13,2	0,52
Organic lake sapropel	3,29	0,04	0,16	1,48	0,22
Siliceous lake sapropel	1,11	0,02	0,55	1,01	0,78
Mixed lake sapropel	2,10	0,16	0,32	1,60	0,30
Manure	1,62	0,40	1,54	0,24	0,17

One of the most important methods of environmental pollution with heavy metals is industrial wasterwater contamination (Kilkus, 1991). Heavy metals accumulate in the sapropel are indicator environmental pollution. Study of heavy metals in sapropel shows that concentrations of metallic elements did not excuded their MPC values for II category sludge presented in LAND 20:2005.

Heavy metals are related to the basic chemical properties of sapropel– moisture, organic matter, carbonate content and elemental composition. Sapropel from the studied lakes is not polluted by industrial sources and can be considered as prospective material for agricultural and other applications (Katkevicius et al., 1998; Stankevica et al, 2012; Paliulis, 2014).

More striking difference could be noticed comparing the heavy metals amounts found in the sapropel of various tipes are, obviously, conditioned by anthropgenic factors of surroundings (Table 3).

Heavy metals concentration depends on the organic matter content of the sediment and the geochemical properties (composition of minerals, rate of their weathering, hydrological and climatological conditions) of the lakes' catchment areas. Increased concentrations of Pb and Cd are found in upper layers of sediments from anthropogenic pollution (Paliulis, 2014).

Table 3. Heavy metals of lakes sapropel

Substances		mg kg^{-1} of sapropel					
		Copper (Cu)	Manganese (Mn)	Iron (Fe)	Lead (Pb)	Chromium (Cr)	Cadmium (Cd)
Calcareous sapropel	lake	8	240	40	17	4	3
Organic sapropel	lake	45	560	12000	16	15	1
Siliceous sapropel	lake	6	70	140	8	3	1
Mixed lake sapropel		2	32	520	2	4	1

Sapropel could be of different colours. Colour is important for assessment of the sapropel. It shows the content of organic and inorganic matter in sapropel. Greenish colour indicates chlorophyll, pink – carotene, blue – vivianite, gray – lime impurities. Black or quickly darkening colour indicates iron (Linčius, 1975).

The most characteristic features of sapropel is its colloidal structure. This structure depends on the contents of iron (sapropel ashes in some cases contains up to 8 % FeO_3), aluminium (Al_2O up to 4 %), silicon (SiO_2 up to 4 %). These mentioned iron, aluminum, silicon oxides form gels. These are colloids with solid structure. They absorb a lot of water – up to 90 %, and are slowly drying out. They evaporate water with difficulty and become completely solid when dry. The researchers point out that even five years after the sapropel insertion into soil, non-degraded sapropel pieces of a few centimeters (up to 3 cm) could still be found. Resulting from its structure, sapropel is characterized by a low filtration.

Organic colloids can absorb a large amount of water. The latter amounts from 40 to 97 % in sapropel. Due to colloids sapropel possesses such features as high water capacity, low filtration and viscosity. These features positively affect soil water capacity, porosity and texture (Katkevicius et al., 1998).

4. Lakes cleaning

In order to restore silted lakes, attempts to get them to their natural state are made. It is a complex process requiring to increase the water depth, improve water quality, eliminate water pollution causes, restore lake vitality and its biocoenosis. The most effective restoration of the lake is its cleaning by removal of the accumulated sludge. The sludge is removed from the lake by hydro-mechanical, mechanical and combined methods (Liuzinas and Jankevicius 2005).

Hydro-mechanical way implies lakes cleaning using dredgers (Ellust 370H; 8PZU-3M; Watermaster Classic III, etc.) (Fig. 4). They consist of buoys, ground pump, suction pipe, pulp pipes. Dredge digs the sludge and in the form of pulp transports it by pulp pipes to the shore into the pre-installed storage places – sedimentation tanks, or spread directly on the fields (Ciūnys and Katkevicius, 2008).

Fig. 4. Dredger in the lake

Mechanical cleaning method is used when there is still a 1.0-2.0 m water layer or when the lake is completely silted. In not completely silted lakes sapropel is excavated using excavator with a grab bucket from buoys. Excavator with a bucket is installed on buoys and thus floats to the sapropel excavating site. Buoys are anchored. Sapropel is excavated with a grab bucket and poured into the barge. The barge moves to the shore where sapropel is scooped or sucked into vehicles and transported to the drying or storage sites (Katkevicius et al., 1998).

In silted lakes sapropel is excavated together with peat using excavators and then loaded into trucks.

Combined method is when both mechanical and hydro-mechanical methods are used. Sludge from the bottom is excavated mechanically and poured into the mixer then

17

through the pipeline it is transported to the shore into storage areas – sedimentation tanks (Ciūnys A. and Katkevičius l. 2008).

5. Sapropel use

Due to its physical properties and chemical composition sapropel can be used widely. Application differs depending upon the sapropel preparation: paste-like, powdery and granular or loose frosted sapropel (Ciūnys et al., 1994).

Sapropel in building materials industry. The structure and chemical composition of sapropel makes it suitable for the use in building materials, ceramics. Already in the sixth decade of the 20th century in researches dealing with application of sapropel for these purposes were started. Adhesive properties of organic sapropel caught the attention; the aim was to at least partially replace the expensive, unhealthy synthetic binders as well as use the sapropel as burnable additive for production of ceramics and agloporit.

In Lithuania, at the Institute of Thermal Insulation, production tests of insulation boards produced from waste and organic sapropel were launched in 1992. It was revealed that the use of sapropel as an adhesive allows the production of thermal insulation panels with a density of 250-320 kg • m-3, flexural strength – 0.1-0.7 MPa and in compression – from 0.3-1.3 MPa. Thermal conductivity of such panels (depending on the density) is 0.052-0.081 W/m • K. Products with sapropel are resistant to molding (Žvironaitė, 1997).

At Kaunas University of Technology effect of various additives, such as foamer, gas producing agents, porous aggregates, burnable materials, on ceramic materials was studied. Sapropels were also tested as burnable components. Previous studies have shown that addition of calcareous sapropel affects the principal technological properties of clay (plasticity, shrinkage, sintering) as well as colour, density, strength, resistance to cold of a ceramic body. It is observed that addition of calcareous sapropel reduces shrinkage of burning samples. Ceramic body kilned from clay supplemented with calcareous sapropel is lighter, with better thermal insulation properties, of brighter colour. It was stated that both the carbon and the organic sapropel can be used as additives in the production of unburned clay materials (Gurskis and Navickas, 2001). At Kaunas University of Technology the use of calcareous and organic sapropel in manufacturing of ceramic products was tested. Dried and pulverized calcareous sapropel was mixed into the clay. It was revealed that samples with addition of 7-10 % of sapropel kiln more rapidly, do not deform while kilning, their strength and hygroscopicity remain almost unchanged. Organic sapropel is suitable for production of porous ceramic, which, according to all standards is fit for manufacturing 38 thermal insulation products. Such ceramic does not change even at 1000 °C temperature. Blending together the additives of organic and calcareous sapropel improves the structure of ceramic products: the decomposition of carbonates releases CO_2, which results in occurrence of huge number of small pores, thus improving the characteristics of the product (Navickas and Gurskas, 2005; Kasperiūnaitė et al., 2010).

Sapropel – a fodder additive. Sapropel contains essential amino acids, that livestock should get with fodder. In lake silt they make up to 47-60 % of free amino acids. Humic acids of sapropel have antiseptic properties (Zaccone et al., 2008), therefore, in order to increase the amount of useful microorganisms, they may be used to stimulate digestive tract of livestock and improve the microbial balance. Owing to its chemical and physical properties, sapropel can be used in livestock feed as vitamin and mineral food supplement. Good results were obtained when the diet of calves was supplemented with sapropel granules. Weight gain of the tested group of calves was 11.5 % higher than of the control group. When sapropel granules were added to the feed of boars, pig fertilization increase was observed (Ciūnys et al., 1994).

In Belarus an extensive research on the use of sapropel granules as supplement to the main livestock rations was performed. Evaluation of carcass showed no change in the animal organism, while the meat yield was slightly higher (Lopotko et al., 1992). So, granules of organic and calcareous sapropel could be used as feed additive for livestock. The compound feed is supplemented with sapropel granules or powder. It extends the variety of available compound feed. In fodder, lime and phosphates are sometimes replaced by sapropel. Sapropel additives increase the biological value of compound feed; enrich it with mineral macro- microelements. In Lithuania attempts to use sapropel for livestock fodder were also made. Already in 1960-1962, B. Malaškaitė carried out the research at the Lithuanian Institute of Animal Husbandry and Veterinary. At the beginning, pigs were fed with 0.5 kg of sapropel a day. After 2 weeks, and then every 2 weeks for 36 weeks the sapropel content was increased by 0.1 kg. Pigs fed with sapropel-enriched fodder required by 2-6 % less fodder units for 1 kg of weight gain; in liver of such pigs the amount of vitamin B_{12} increased almost twice. When birds were fed with food containing up to 80 % of sapropel, their weight by 11-22 % exceeded the weight of birds fed the same food but without sapropel (Dudin et al., 1997).

Feeding sapropel is useful supplement for animals and poultry, it contains a set of biologically active substances increasing pigs, cattle stock and poultry's productivity, and it improves resistance to diseases of different kinds. Feeding sapropel was designed as a supplement for all kinds of animals and poultry's diet naturally with humidity not more than 65 %. Dry feeding sapropel may be used in compound feed industry for all kinds of combined feed production as an organic and mineral supplement (premix) and as filler for vitamin premixes (Valius et al., 1962; Mikulioniene and Balezentiene, 2009).

Sapropel is used as a supplement not only for feeding stuff reduction, but also for the purpose of diet beneficiation with mineral materials, amino acids, vitamins, and biologically active materials. Sapropel represents valuable vitamin and mineral supplement (Kanopkaite et al., 1962).

Replacement of sapropel's part provides nutrient budget cheapening for 7 – 10 % and in case of 10% replacement provides poultry broiler meat production profitability on a level of control group and expensive nutrition ingredients saving. According to economic indicators additional usage of sapropel with humidity of 60 % added to the main ration is more efficient - profit of 11,18 % over control level was received from the average usage (Lopotko et al., 1992; Mikulionieme et al., 1998).

Sapropel usage efficiency at broiler-chicken feeding depends on feeding method (feed to appetite, part replacement, feed mix applicator), on its dosage added into ration and on sapropel humidity. Sapropel usage influences on feed mix eatability, digestibility and nutrition, growth rate, meat productivity and poultry meat production economic indicators. Sapropel added in broiler diet decreases feedstuff general cost for 7,24 - 17,74 % in experimental groups vs. control groups (Mikulioniene et al., 1998).

Sapropel for mud baths and cosmetics. Sapropel is ecologically the cleanest material, with no toxic substances, contamination with heavy metals or radionuclides. The main ingredient is organogenic substrate, formed only in certain lakes, exceptional natural habitats, under the influence of particular natural phenomena. Sapropel is characterized by antiradiation properties. Studies have shown that sapropel as much as 20 times reduces the radionuclide mobility; it is therefore suitable for balneology, medical purposes, cosmetics (Tarcijonaitė and Miknevičius, 2014). Application of this product gently cleanses the skin, removes dead cells, activates blood circulation, skin cells acquire more elasticity, general wellbeing improves. Substances present in sapropel possess antibacterial effects, so during the treatment 95 % of pathogenic bacteria, fungi and toxins are destroyed. Black therapeutic mud is recommended for people with oily skin problems, because it is rich in organic matter, various ions and trace elements. Infusive force of nature is contained in therapeutic muds. Yet ancient Egyptians used mud's healing properties for cure, cosmetic purposes and for organism youth renewal. Recently muds of the Dead Sea, Spain and Hungary became very popular. However not all have an opportunity to recover their health not only at international but also at local resorts (Samutin, 1997).

Sapropel muds composition is complicated. Its basis is composed of calcium, magnesium, ferrum, zinc, selenium and manganese salines; aluminium and silicium colloidal complexes, up to 50 microelements, amino acids, eicosanoic acids, phospholipids, groups of B, A, D, E vitamins, folic acid, antioxidants, hormons-like agents, antibiotic-like agents, microorganisms physiological group, carotenoids, bitumens, lignin, cellulose, phenolics and flavonoids. Sapropel acid-base balance (pH) level is close to human ideal pH (5.5 - 6.05) Lopotko and Jevdokimova, 1986; Katkevičius et al., 1998).

Sapropel mud is fine mass with particulates size up to 25 micron, which allows them to penetrate deeply through skin sweat and oil glands. Optimal interchangeable process is taken place: an organism ingests sapropel healing properties returning accumulated toxins and poisons often appearing as a reason of various skin and allergic diseases. Besides chemical and thermal influence factor also there is informational influence of sapropel on an organism by means of accumulated energy of primeval nature. This energy has been accumulated in sapropel muds during millenniums, therefore their healing action is appreciated even after first procedures (medicalency.com/grjazelechenie.htm.)

Sapropel for soil fertilization. Studies on chemical composition of sapropel show that it contains plenty of components that positively affect soil fertility. The sapropel with ash content less than 50 % is the most appropriate for preparation of fertilizers. However, more mineralized sapropel can also be used for soil fertilization, because together with organic matter the soils are enriched with macro and micro elements. Calcareous sapropel is suitable for soil liming. The value of sapropel as a fertilizer has long been known. (Baksiene et al., 2006; Bakksiene, 2008; Baksiene and Ciūnys, 2012; Baksiene and Asakaviciute, 2013).

6. Preparation of sapropel as a fertilizer

Since sapropel accumulates in lakes under anaerobic conditions, toxins can be identified only in extracted sapropel. In order to remove the toxins it is recomended to keep sapropel in sediment bowls for sometime prior to use. However, nitrogen content tends to decline in stored and dewatered sapropel (Chochlova, 1997; James et al., 2003).

Sapropel excavated from lakes contains a large amount of water (85-95 %). For its removal, sapropel by pulp pipes is transported into sedimentation tanks (Fig. 5).

Fig. 5. Mound sedimentation tanks set up in the field
(photo by A. Ciūnys)

Surface incline of organic silt is smaller, while of mineral and calcareous sapropels – larger. In some cases, silt layer exceeding 1.0 m could be poured into sedimentation tanks. However, the fact that the higher are the sedimentation mounds, the more useful space they occupy, should be considered. If the mound height is 1.0 m, the useful area of the sedimentation tank is smaller. Poured into sedimentation tanks the silt is left for frosting throughout the winter. Before winter, the silt dries and layer thickness reduces to 0.7-0.8 m. The surface layer cracks into pieces (Fig. 6).

Fig. 6. Crumbled peat layer in sedimentation tanks (photo by A. Ciūnys)

In winter the depth of freezing is about 0.3 m. Basically separate frosted silt clumps form. Therefore, for deeper layers of silt to frost, it is advisable to remove the top layer of frosted silt (separate silt clumps) so that until spring deeper layers could also frost (Fig. 7). This way the frosting of the entire silt is achieved. In spring the frosted silt thaws, separate clumps crumble, quickly dry, silt becomes loose. Such loose silt is manually poured into bags and delivered to trade networks.

Fig. 7. Freezing sapropel

When silt is shoved by mechanical means (bulldozers, loaders), certain amount of base mineral soil mixes together, so the silt loses its quality and appearance. Silt shoved by bulldozed into piles is later suitable only for spreading in the fields.

Sapropel could be dehydrated employing macrophytes. For this purpose sedimentation tanks are filled with 45 cm thick sapropel layer. Its small particles are allowed to settle for 3-5 days. Then water plants – macrophytes, are planted into the compacted substance. Tests showed that simple reed (*Phragmites australis*), common cattail (*Typha latifolia*), sweet flag (*Acorus calamus*) and three-lobe beggarticks (*Bidens tripartita*) are most suitable for the purpose (Liuzinas and Jankevicius, 2005).

In Latvia sapropel is raised from lakes by hydraulic means and stored in reservoirs. Then, retaining its natural moisture, the sapropel is mixed with peat.

In some cases powdery or granulated sapropel is prepared. Sapropel, dried to 60-70 % moisture, is transported into the factory and passed through special granule-producing equipment. The produced granules are transported for further drying. AVM-1.5 dryer with a slanted drum is most commonly used. Hot air (80-90 °C) is blown through the drum, and granules are dried to 15-18 %.

Powdery sapropel is produced when sapropel pulp is sprayed into the drying drum together with hot air (about 200 °C). Hot temperature does not impair the sapropel quality.

Organic Fertilizers Based on Sapropel and Peat. Application of organic-mineral fertilizers (OMF) based on sapropel and peat is of considerable interest. These include the products called "Humin Plus", "Sapropeat", "Sapropeat Uni", "Sapro Agro", and "Sapro Elixir". Sapro Agro (manufactured by LLC LATPOWER, Riga, Latvia) is biologically active soil conditioner, produced on environmentally sound technology from natural ingredients: sapropel colloid and active peat (http://www.latpower.lv/index.php?page=2). Sapro Elixir (manufactured by SIA HUMIN VIT, Ogre, Latvia) is produced from natural fresh water lake sapropel with natural moisture (Ostrovskij, 2014; Ostrovskij et al., 2014). Sapro Elixir contains a full spectrum of natural biologically active ingredients. The product is balanced with limited amounts of NPK and nutrients (http://www.huminvit.lv/sapro-elixir?lang=eng). Sapro Elixir is a highly-efficient natural-organic fertilizer and soil enhancer that can be applied on all types of soils and all kinds of fruit and vegetables, sowing and decorative crops, trees, and bushes. Humin Plus/Sapropeet (manufactured by the German-Ukrainian Center for Innovative Agri-industrial Technologies – FuTech) is an organic mineral micro-fertilizer based on a sapropel extract. A licensed technology based on new physical principles in the processing of raw materials (cavitation combined with magnetic treatment) is used to obtain micro fertilizers with new characteristics (i.e.: improved consistency and increased physiological and biological activity of the ingredients) (http://athdevelopments.co.uk/downloads-/Humin%20Plus%20Brochure.pdf). Sapropeat Uni is a biologically active product aimed at rectifying soil health.

7. Lakes sapropel for soil fertilization

Fertilization with lakes sapropel enriches the soil with organic matter as well as improves its atructure, physical and microbiological properties; it is also an inportant measure preventing wind erosion of soil. Practically all investigatiors agree that lake sediments function as long-term measure for improving agrochemical and physical properties of soil. The effect of soil sediments highly increases if they are composted with manure. The amount of microorganisms in such sediments raises and consequently does the amount of nutritious substances.

In the works of Norwegian and Canadian scientists (Sveistrup et. al., 1995, Zebarth et. al., 1999) the conclusions about positive effect of lake sediments on physical properties of soil. Introduction of sediments increased the soil porosity and moisture retention capacity of soil, improved the soil texture and quality. In Voke Branch of the Lithuanian Institute of Agriculture investigation concerning the application of lake sediments for soil fertilisation were starte din 1994. The aims of the research is to establish the influence of lake sediments and their mixtures with other organic matter (manure, sewage) and mineral NPK fertilisers upon the yield, agrochemical and physical properties of soil; to compare the effect the of sediments with that of the sediments-manure mixture.

Investigations carried out in France, Italy, Russian and United Kingdom countries on the effects of lake sediments allow us to suppose that its impact depends upon its chemical composition (Grishina et al. 1990; Andresini et al. 2003). Calcareous lake sediments are more suitable as a measure for soil liming, organic and siliceous ones are suitable as a source of nutritious materials (Orlov and Sadovnikova 1996; Roberts et al. 2011).

8. Application of lakes sapropel for improvement of soil chemical properties

In Lithuania the efficiency of calcareous lake sediments has been investigated extensively. In sandy loam soil and sandy soil in crop rotation fields the rates of 50, 100, 150, and 200 t ha^{-1} (of dry matter) of calcareous sediments from Lake Ilgutis (Vilnius district) were investigated we tests various rates of sediments functioned as long-term measure of improved agrochemical and physical properties of soil. After fertilization with lake sediments, acidity of soil decreased, content of humus increased, and qualitative composition improved. All rates of sediments improved soil texture and moisture regime. However, various rates of calcareous sediments had no positive effect upon the productivity of crop rotation. The research results showed that the application of sediments in sandy loam soil (pH 6.0) in the 1st crop rotation season increased the yield of agricultural plants by 2–5 %. Only after application of the largest rate (– 200 t ha$^{-1)}$ of dry sediments – the productivity increased by 7 % during the 2nd crop rotation season. It may be predicated that sediments containing larger amount of organic material is more effective for the yield of agricultural plants (Baksiene and Janusiene 2005).

According to Russian scientists, organic lake sediments are most effective. Having carried out field experiments they have determined that organic lake sediments were no less efficient than peat-manure compost in sandy loam soil, sometimes even superior. The rate of 60–80 t ha^{-1} of sediments produced the additional yield of 0.34–1.61 t ha^{-1} of barley, while the same rate of peat-manure compost produced the additional yield of 0.28–1.06 t ha^{-1} (Grigorov and Ovchinnikov 1994; Ostrovskij et al., 2014). Practically all investigators agree that lake sediments function as long-term measure for improving agrochemical and physical properties of soil (Booth et al. 2007).

8.1. Indices of soil acidity

Optimal soil acidity is a very important indicator which determines functionality and productivity of nutrients in soil. In Lithuania, about 40% of soils have increased acidity (pH $\leq$ 5.5). After an intensive liming the area of soils with high acidity has reduced by 18.6 % (Mazvila et al., 2006). Liming is one of the most important and efficient ways of reducing acidity in soils, therefore making them better. For this purpose, not only limestone (which is commonly used) but also local lime fertilizers could be used. The experiments carried out in Vezaiciai Branch of Lithuanian Research Centre for Agriculture and Forestry revealed that the fastest way to neutralize soil was to use slaked lime, but its activity lasted for the shortest period of time. Other local lime fertilizers like mud, limestone tufa and calcareous loam were active for a longer period of time and they were more effective for yield (Ozeraitiene et al., 2006). Most of the local lime fertilizers are found in freshwater bodies calcareous sediments. In literature, these fertilizers are called differently lake lime and marl, tufa, etc. Calcareous sapropel belongs to this category (Rubinstein, 1984). Various investigations of sapropel were carried out in Latvia, Russia and Byelorussia (Chochlova, 1997, Kireycheva and Chokhlova, 2004, Liepins, 1995, Lopotko, 1992, Orlov and Sadovnikova, 1996). The data of these researches shows that all kinds of sapropel (with different chemical composition) are adequate for fertilizing wealds. They are especially effective in light texture granulometric soils.

After application of calcareous sapropel, limestone or its mixtures for soil fertilization on the background without mineral fertilizers, indicators of soil acidity had changed after the first crop rotation. Slightly neutral soil became neutral (Fig. 8). Soil pH and total exchangeable bases increased by 0.9-1.2 units and 10.7-86.8 meq kg-1 soil, respectively. After fertilization with manure, total exchangeable bases increased by 11.6 meq kg-1 soil, but according the pH indicator, soil from slightly neutral (pH-6.1) became acid (pH-5.7).

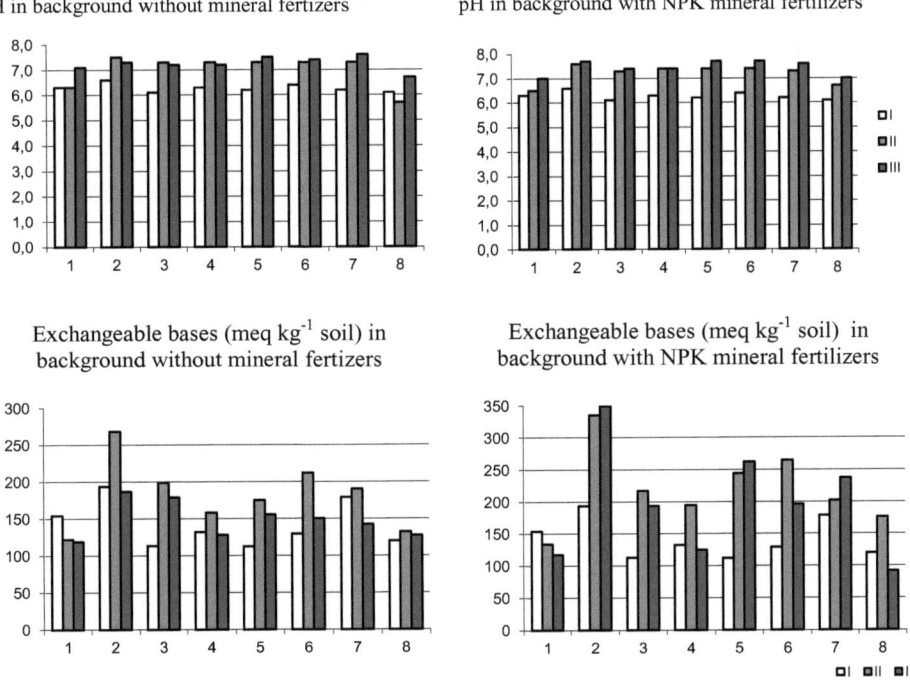

Fig. 8. Influence of calcareous sapropel and limestone on the indices of sandy loamy Cambisol acidity

I – acidity indices before trials; II – after first crop rotation (6 year); III – after second crop rotation (11 year). 1-8 - treatments: 1) control; 2) 10 t ha^{-1} limestone (CaCO3); 3) 25 t ha^{-1} sapropel (S); 4) 1.5 rate of CaCO3 + 25 tha^{-1} manure (M); 5) 25 t ha^{-1} sapropel (S) +10 t ha^{-1} manure (M); 6) 25 t ha^{-1} sapropel (S) + 25 t ha^{-1} manure (M); 7) 25 t ha^{-1} sapropel (S) + 10 m^{3}sewage (Sw); 8) 65 tha-1 manure (M).

After the second rotation, pH indicator remained nearly the same (7.3-7.5) but total exchangeable bases had markedly decreased. After fertilization with calcareous sapropel, total exchangeable bases decreased but did not reach the level of primary indicator (before start of the experiments).

After fertilization of soil with limestone, this indicator was lower than before the start of the experiments (decreased from 194.3-131.6 to 187.3-127.6 meq kg^{-1} soil).

On the background with minimal rates of mineral fertilizers, application of calcareous sapropel and limestone and its mixtures with manure also decreased soil acidity, increased pH by 1.0-1.2 and total exchangeable bases by 23.0-140.6 meq kg-1 soil. After the second crop rotation, differently from the background without mineral fertilizers, total exchangeable bases decreased significantly (till 92.2 meq kg-1 soil) only in the eighth treatment, where 65 t ha-1 manure had been used for fertilizing.

Since soil acidity was neutral before the establishment of experimental plots and the amount of calcium was low in the investigated organic sapropel the soil acidity did not change, after 11 years of crop rotation, on the background without mineral fertilizers– pH remained almost the same (Fig. 9). The exchangeable bases also increased from 104.8 to 228.8 meq kg^{-1} of soil. Analogous investigations were performed in Byelorussia. The results showed that calcareous sapropel outfaced lime and limestone, and in some cases, it was even better because it enriched soil with nutritive elements that are needed for plants to grow better (Lopotko et al., 1992).

On the background with mineral fertilizers, when soil was fertilized with various rates of sapropel, mixtures of 10 t ha^{-1} sapropel plus 25 t ha^{-1} manure, and 10 t ha^{-1} sediments plus 10 m^3 sewage, the soil acidity changed from rather neutral into rather acid. Soil pH changed from 6.5 to 5.7, the exchangeable bases decreased from 108.4 to 97.0 meq kg^{-1}. After fertilization with manure on both backgrounds soil pH and the exchangeable bases of soil increased by 84.9–92.6 meq kg^{-1}.

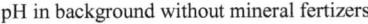

pH in background without mineral fertizers

pH in background with NPK mineral fertilizers

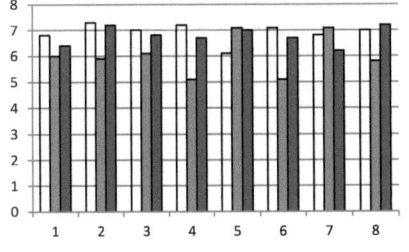

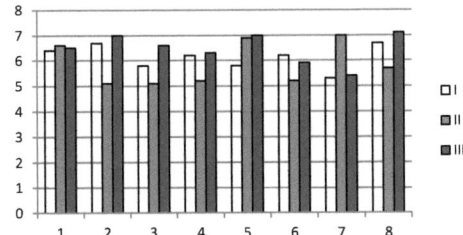

Exchangeable bases (meq kg^{-1} soil) in background without mineral fertizers

Exchangeable bases (meq kg^{-1} soil) in background with NPK mineral fertilizers

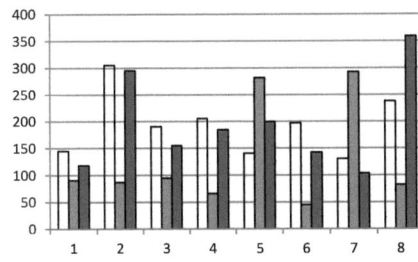

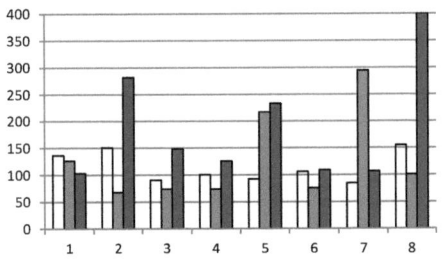

Fig. 9. Influence of organic sapropel on the indices of sandy loamy Cambisol acidity
I – acidity indices before trials; II – after first crop rotation (6 year); III – after second crop rotation (11 year).
1-8 - treatments: 1) Control; 2) 10 t ha^{-1} sapropel (S); 3) 20 t ha^{-1} sapropel; (S); 4) 40 t ha^{-1} sapropel (S); 5) 10 t ha^{-1} sapropel (S) + 10 t ha^{-1} manure (M); 6) 10 t ha^{-1} sapropel (S) + 25 t ha^{-1} manure (M); 7) 10 t ha^{-1} sapropel (S) + 10 m^{3}ha^{-1} sewage (Sw); 8) 65 t ha^{-1} manure (M).

Although 1.01% of calcium was identified in siliceous lake sapropel, it caused no significant changes in soil acidity (Fig. 10). According to pH, soil acidity remained neutral. The total exchangeable bases changed within the standard error. In case the sediment-manure mixture and sewage mixtures were applied, neutral soil became slightly acidic (pH value 6.0), and in case manure was used it turned more acidic (pH value 5.5). Consequently, in these treatments, the total exchangeable bases decreased by 17.3 to 56.1 meq kg^{-1} .

pH in background without mineral fertizers

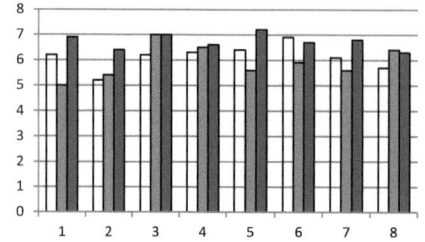

pH in background with NPK mineral fertilizers

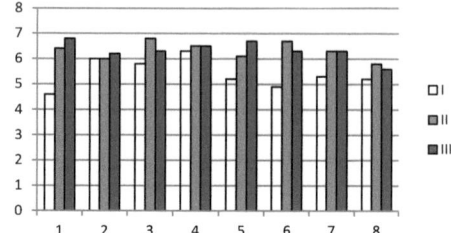

Exchangeable bases (meq kg^{-1} soil) in background without mineral fertizers

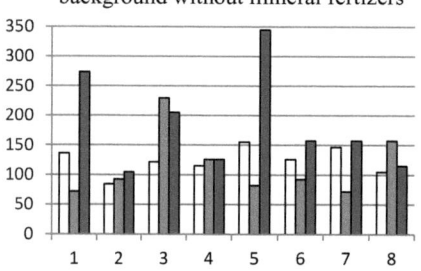

Exchangeable bases (meq kg^{-1} soil) in background with NPK mineral fertilizers

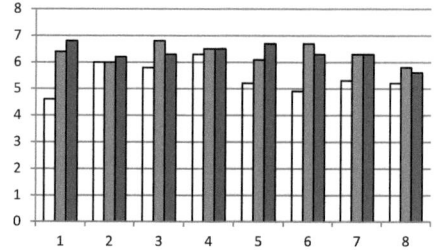

Fig. 10. Influence of siliceous sapropel on the indices of sandy loamy Cambisol acidity

I – acidity indices before trials; II – after first crop rotation (6 year); III – after second crop rotation (11 year). 1-8 - treatments: 1) Control; 2) 25 t ha^{-1} sapropel (S); 3) 50 t ha^{-1} sapropel (S); 4) 100 t ha^{-1} sapropel (S); 5) 25 t ha^{-1} sapropel (S) + 10 t ha^{-1} manure (M); 6) 25 t ha^{-1} sapropel (S) + 25 t ha^{-1} manure (M); 7) 25 t ha^{-1} sapropel (S) + 10 m^{3}ha^{-1} sewage (Sw); 8) 65 t ha^{-1} manure (M).

8.2. Total nitrogen

Total amount of nitrogen after calcareous sapropel treatments did not change significantly but it decreased somewhat more (0.011%) after liming and mixture of sapropel with 10 t ha-1 manure (0.028%) and sewage (0.012%) (Fig. 11). The amount of nitrogen increased by 0.011% after fertilization with 25 t ha-1 sapropel and 25 t ha-1 manure mixture. At the end of the first rotation, the biggest amount of total nitrogen was found in soil which had been fertilized with mixture of sapropel and manure.

Total N in background without mineral fertizers   Total N in background with NPK mineral fertilizers

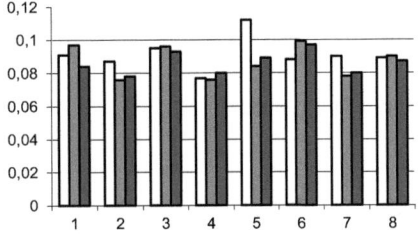
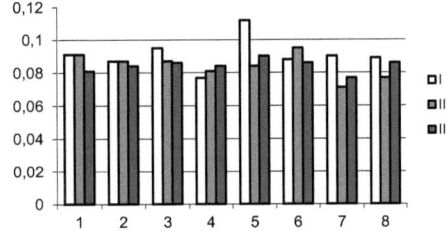

Fig. 11. Influence of calcareous sapropel and limestone on the indices of sandy loamy Cambisol total nitrogen
I – total nitrogen indices before trials; II – after first crop rotation (6 year); III – after second crop rotation (11 year). 1-8 - treatments: 1) control; 2) 10 t ha^{-1} limestone (CaCO3); 3) 25 t ha^{-1} sapropel (S); 4) 1.5 rate of CaCO3 + 25 tha^{-1} manure (M); 5) 25 t ha^{-1} sapropel (S) +10 t ha^{-1} manure (M); 6) 25 t ha^{-1} sapropel (S) + 25 t ha^{-1} manure (M); 7) 25 t ha^{-1} sapropel (S) + 10 m^{3}sewage (Sw); 8) 65 tha-1 manure (M).

In the background with mineral fertilizers the amount of total nitrogen was lower (0.013%) in the treatment where manure had been used for fertilizing. In other cases, the amount of total nitrogen did not markedly change.

Fertilization of soil with various rates of lake organic sapropel, its mixtures and manure influenced a little bit the amount of total nitrogen (Fig. 12).

Total N in background without mineral fertizers   Total N in background with NPK mineral fertilizers

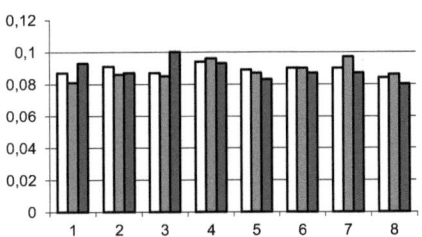
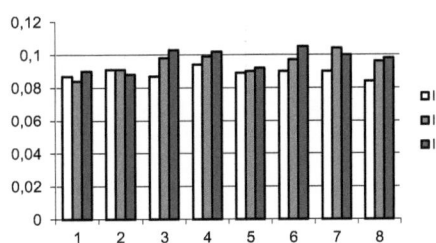

Fig. 12. Influence of organic sapropel on the indices of sandy loamy Cambisol total nitrogen
I – total nitrogen indices before trials; II – after first crop rotation (6 year); III – after second crop rotation (11 year). 1-8 - treatments: 1) Control; 2) 10 t ha^{-1} sapropel (S); 3) 20 t ha^{-1} sapropel; (S); 4) 40 t ha^{-1} sapropel (S); 5) 10 t ha^{-1} sapropel (S) + 10 t ha^{-1} manure (M); 6) 10 t ha^{-1} sapropel (S) + 25 t ha^{-1} manure (M); 7) 10 t ha^{-1} sapropel (S) + 10 m^{3}ha^{-1} sewage (Sw); 8) 65 t ha^{-1} manure (M).

The highest amounts of nitrogen were introduced into the soil with various rates of lake siliceous sapropel and sapropel-manure mixtures. During the eleven years of crop rotation, plants did not exhaust the amounts. Therefore, the amount of total nitrogen in the soil was by 0.009–0.021 percentage units higher (Fig. 13).

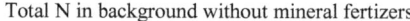

Total N in background without mineral fertizers   Total N in background with NPK mineral fertilizers

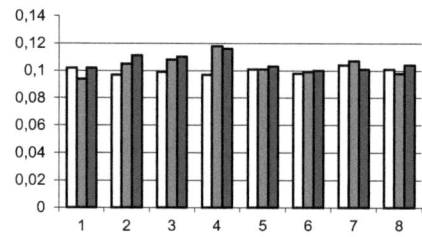
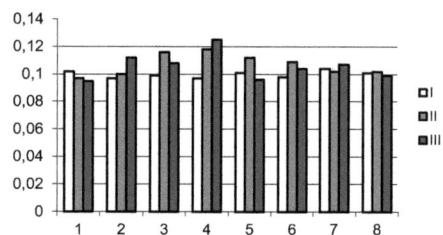

Fig. 13. Influence of siliceous sapropel on the indices of sandy loamy Cambisol total nitrogen
I – total nitrogen indices before trials; II – after first crop rotation (6 year); III – after second crop rotation (11 year). 1-8 - treatments: 1) Control; 2) 25 t ha^{-1} sapropel (S); 3) 50 t ha^{-1} sapropel (S); 4) 100 t ha^{-1} sapropel (S); 5) 25 t ha^{-1} sapropel (S) + 10 t ha^{-1} manure (M); 6) 25 t ha^{-1} sapropel (S) + 25 t ha^{-1} manure (M); 7) 25 t ha^{-1} sapropel (S) + 10 m^{3}ha^{-1} sewage (Sw); 8) 65 t ha^{-1} manure (M).

Gradually increased rates of lake sapropel enhanced the amount of total nitrogen from 0.097 to 0.118% in the soil. Fertilization with manure had no impact on the amount of nitrogen. After fertilization with lake sediments at 100 t ha-1 the content of total nitrogen increased by 0.002–0.021 and that of humus by 0.53 percentage units.

8.3. Mobile phosphorus and potassium

Low amounts of phosphorus and potassium were found in calcareous sapropel. Not much of the substances were inserted into the soil. That is why, in all fertilization treatments after the first and second rotations phosphorus in soil has decreased by 2.2-41.4 mg kg^{-1} and by 11.5-42.5 mg kg^{-1} (Fig. 14).

After the first and second crop rotations, minimal rates of mineral P and K fertilizers increased the amount of phosphorus by 41.5-94.0 mg kg-1 soil and amount of potassium by 10.0-30.7 mg kg-1 in all fertilization treatments

Mobile P in background without mineral fertizers

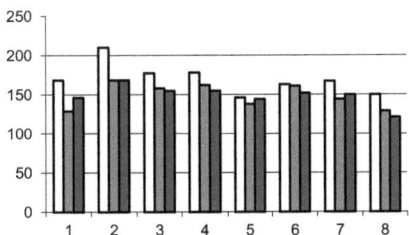

Mobile P in background with NPK mineral fertilizers

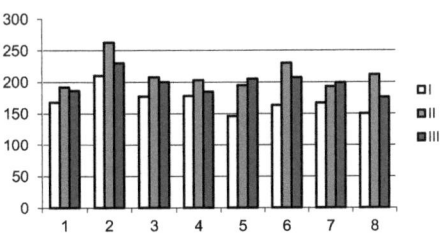

Mobile K in background without mineral fertizers

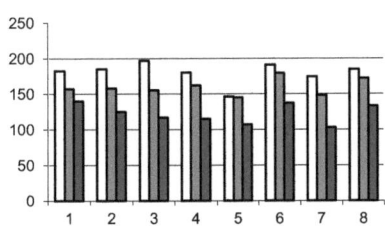

Mobile K in background with NPK mineral fertilizers

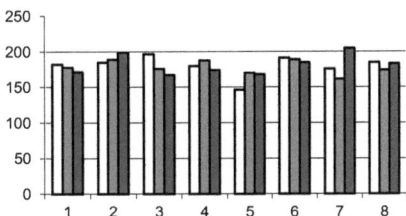

Fig. 14. Influence of calcareous sapropel and limestone on the indices of sandy loamy Cambisol mobile phosphorus and potassium

I – total nitrogen indices before trials; II – after first crop rotation (6 year); III – after second crop rotation (11 year). 1-8 - treatments: 1) control; 2) 10 t ha^{-1} limestone (CaCO3); 3) 25 t ha^{-1} sapropel (S); 4) 1.5 rate of CaCO3 + 25 tha^{-1} sapropel (S) + 10 m^{3}sewage (Sw); 8) 65 tha-1 manure (M). manure (M); 5) 25 t ha^{-1} sapropel (S) +10 t ha^{-1} manure (M); 6) 25 t ha^{-1} sapropel (S) + 25 t ha^{-1} manure (M); 7) 25 t ha^{-1}

Small amount of phosphorus (0.04 %) was detected in organic sediments, therefore a little amount of phosphorus reached the soil. Fertilization with mineral (non-organic) fertilizers made a larger effect on the alterations of mobile phosphorus. On the background without mineral fertilizers in almost all treatments the amount of phosphorus was smaller by 7.9–23.2 mg kg^{-1} to compare with its amount before the establishment of experimental plots; on the background with mineral fertilizers in almost all variants it was larger by 13.2–32.8 mg kg^{-1}(Fig. 15).

Mobile P in background without mineral fertizers

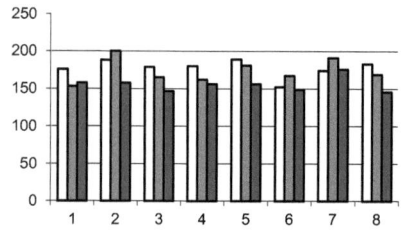

Mobile P in background with NPK mineral fertilizers

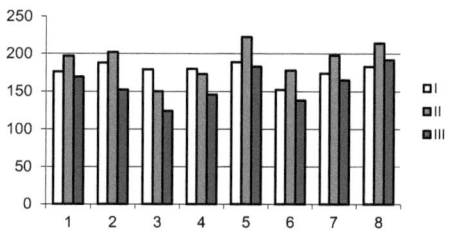

Mobile K in background without mineral fertizers

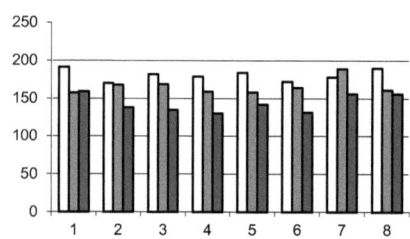

Mobile K in background with NPK mineral fertilizers

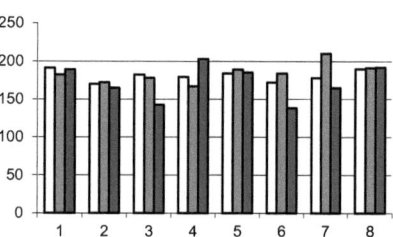

Fig. 15. Influence of organic sapropel on the indices of sandy loamy Cambisol mobile phosphorus and potassium

I – acidity indices before trials; II – after first crop rotation (6 year); III – after second crop rotation (11 year). 1-8 - treatments: 1) Control; 2) 10 t ha^{-1} sapropel (S); 3) 20 t ha^{-1} sapropel; (S); 4) 40 t ha^{-1} sapropel (S); 5) 10 t ha^{-1} sapropel (S) + 10 t ha^{-1} manure (M); 6) 10 t ha^{-1} sapropel (S) + 25 t ha^{-1} manure (M); 7) 10 t ha^{-1} sapropel (S) + 10 m^{3}ha^{-1} sewage (Sw); 8) 65 t ha^{-1} manure (M).

Bigger amount of phosphorus was detected in soil previously fertilized with manure (214.5 mg kg^{-1} of soil) and with sediments -manure mixture (178.3–198.0 mg kg^{-1}of soil).

Similarly, the amount of potassium in organic sediments was rather low, too. Therefore, due to various ways of fertilization the amount of active potassium in soil changed similarly to that of phosphorus. On the background without mineral fertilizers (when organic fertilisers were applied) the defined amount of potassium (2.0–28.8 mg kg^{-1} of soil) was lower almost in all treatments than its amount before the establishment of an experimental plot. However, on the background with mineral fertilizers it was larger (2.0–32.3 mg kg^{-1} of soil) almost in all treatments. Different fertilization with organic fertilizers did not influence the amount of potassium in soil.

Larger amounts of phosphorus were introduced into soil with higher rates of lake siliciceous sapropel (50, 100 t ha^{-1}), sapropel-manure mixture, and pure manure (163,7-196,0 mg kg^{-1}) (Fig. 16). On the beckground with mineral fertilizers the amount of mobile phosphorus increased by 1,7-54,3 mg kg^{-1}.

The application of siliceous sapropel on the background without mineral fertilizers the amount of mobile potassium decreased by 6,8-43,1 mg kg^{-1} in all treatments of organic fertilization. Amount of potassium, sufficient for plants, was compensated only by minimal rates of mineral fertilizers.

Mobile P in background without mineral fertizers

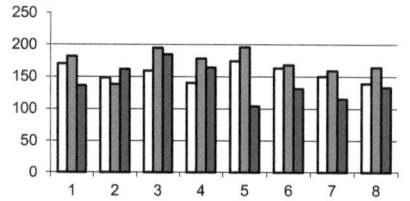

Mobile P in background with NPK mineral fertilizers

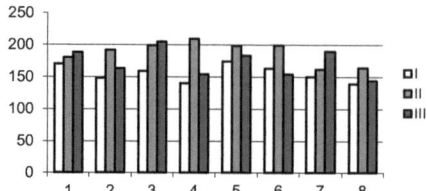

Mobile K in background without mineral fertizers

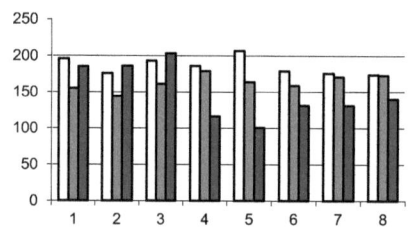

Mobile K in background with NPK mineral fertilizers

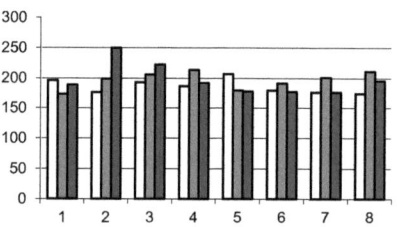

Fig. 16. Influence of siliceous sapropel on the indices of sandy loamy Cambisol mobile phosphorus and potassium

I – total nitrogen indices before trials; II – after first crop rotation (6 year); III – after second crop rotation (11 year). 1-8 - treatments: 1) Control; 2) 25 t ha^{-1} sapropel (S); 3) 50 t ha^{-1} sapropel (S); 4) 100 t ha^{-1} sapropel (S); 5) 25 t ha^{-1} sapropel (S) + 10 t ha^{-1} manure (M); 6) 25 t ha^{-1} sapropel (S) + 25 t ha^{-1} manure (M); 7) 25 t ha^{-1} sapropel (S) + 10 m^{3}ha^{-1} sewage (Sw); 8) 65 t ha^{-1} manure (M).

8.4. Humus and quality

Comparison of the data of humus composition in the treatments applied with calcareous sapropel with the data in the treatments applied with manure suggests that the content of mobile humus substances reduced in the treatments with the higher rates (100-200 t/ha) of sapropel. Calcareous sapropel applied in the soil increased the amount of stable humus substance (Fig. 17).

Humus content in background without mineral fertizers

Humus content in background with NPK mineral fertilizers

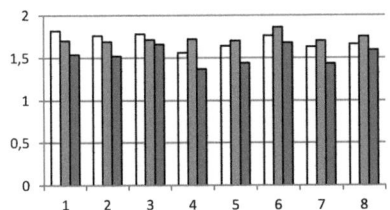 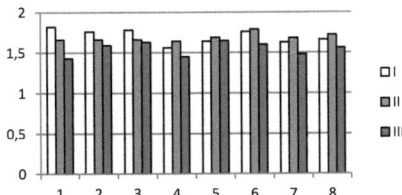

Fig. 17. Influence of calcareous sapropel and limestone on the indices of sandy loamy Cambisol humus content.
I – total nitrogen indices before trials; II – after first crop rotation (6 year); III – after second crop rotation (11 year). 1-8 - treatments: 1) Control; 2) 10 t ha⁻¹ sapropel (S); 3) 20 t ha⁻¹ sapropel; (S); 4) 40 t ha⁻¹ sapropel (S); 5) 10 t ha⁻¹ sapropel (S) + 10 t ha⁻¹ manure (M); 6) 10 t ha⁻¹ sapropel (S) + 25 t ha⁻¹ manure (M); 7) 10 t ha⁻¹ sapropel (S) + 10 m³ha⁻¹ sewage (Sw); 8) 65 t ha⁻¹ manure (M).

Experimental evidence suggests that calcareous sapropel as the lime-containing material not only reduces soil acidity but also improves and stabilizes soil properties necessary for plant growth.

The introduction of different rates of dry sapropel in a sandy loam had influence on the increase of crop productivity in three crop rotations.

On the background without mineral fertilizers higher content of humus (1.71 %) was found in soil fertilized with the highest rate of sediments. In other fertilization treatments higher percentage of humus was detected due to the impact of mineral nitrogen.

Organic sapropel affected the accumulation of humus in soil. Proportionally increased rates of sediments evenly increased the amount of humus from 0.09 to 0.43% (Fig. 18).

Humus content in background without mineral fertizers

Humus content in background with NPK mineral fertilizers

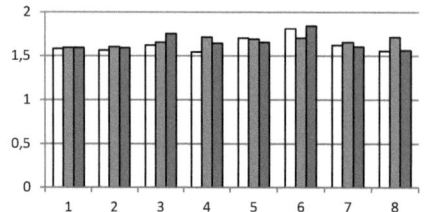
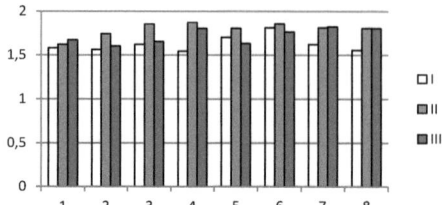

Fig. 18. Influence of organic sapropel on the indices of sandy loamy Cambisol total nitrogen

I – total nitrogen indices before trials; II – after first crop rotation (6 year); III – after second crop rotation (11 year). 1-8 - treatments: 1) Control; 2) 10 t ha^{-1} sapropel (S); 3) 20 t ha^{-1} sapropel; (S); 4) 40 t ha^{-1} sapropel (S); 5) 10 t ha^{-1} sapropel (S) + 10 t ha^{-1} manure (M); 6) 10 t ha^{-1} sapropel (S) + 25 t ha^{-1} manure (M); 7) 10 t ha^{-1} sapropel (S) + 10 m^{3}ha^{-1} sewage (Sw); 8) 65 t ha^{-1} manure (M).

Mixtures of sediments and manure, and pure manure with mineral fertilizers produced a weaker effect. Under their impact the amount of humus in the soil increased by 0.09–0.17%. Larger amounts of phosphorus were introduced into the soil with higher rates of lake sediments (50, 100 t ha-1), sediment-manure mixture, and pure manure (163.7–196.0 mg kg-1).

The amount of humus after fertilization with siliceous sapropel hardly changed, just like of total nitrogen but its changes were not similar to those of nitrogen. A slightly higher amount of humus was found in treatments where bigger rates of manure had been applied in mixtures. Here humus increased by 0.16, 0.10 and 0.09 % (Fig. 19). Analysis of soil performed in 2010 revealed that the amount of humus had decreased in all fertilization treatments and was lower than before the start of the investigations. Compared with the background without mineral fertilizers the amount of humus decreased very slightly (0.01-0.08%). Such decrease was noticed in all fertilizing treatments of the investigation. After the second crop rotation, total nitrogen, similarly as humus decreased in all fertilizing treatments. Minimal rates of mineral nitrogen fertilizer were not sufficient to fertilize soil.

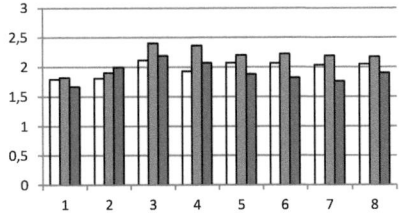

Humus content in background without mineral fertizers

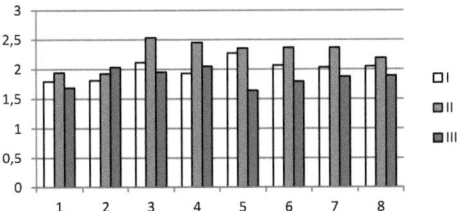

Humus content in background with NPK mineral fertilizers

□ I
□ II
□ III

Fig. 19. Influence of siliceous sapropel on the indices of sandy loamy Cambisol humus content.

I – total nitrogen indices before trials; II – after first crop rotation (6 year); III – after second crop rotation (11 year). 1-8 - treatments: 1) Control; 2) 25 t ha⁻¹ sapropel (S); 3) 50 t ha⁻¹ sapropel (S); 4) 100 t ha⁻¹ sapropel (S); 5) 25 t ha⁻¹ sapropel (S) + 10 t ha⁻¹ manure (M); 6) 25 t ha⁻¹ sapropel (S) + 25 t ha⁻¹ manure (M); 7) 25 t ha⁻¹ sapropel (S) + 10 m³ha⁻¹ sewage (Sw); 8) 65 t ha⁻¹ manure (M).

Organic carbon content accounted for 14.8±0.10% in the composition of calcareous sapropel used in our tests. The larger part of total organic carbon of sapropel was accumulated in insoluble residue (75.6±0.61% from C_{org}), less – in humic acids (9.6±0.3% from C_{org} and fulvic acids (14.8±0.20% from C_{org}). Humic acids composition was dominated by humic acids strongly bound with clay minerals (63.5% from total HA), less humic acids bound with calcum (27.1% from total HA) and still less mobile humic acids (9.4% from total HA). Consequently, humus substances incorporated onto the soil with calcareous sapropel were stable in terms of decomposition. Humus substances in the soil are gradually hydrolysed, and the regenerated humic acids are analogous to the acids present inthe soil and take part in organic matter metabolism (Stepanova and Orlov, 1996). Tests done by radio carbonic method have shown that on light-textured soil humus readily mineralises and more than a half of its composition goes for hydrolysed part (Cherkinsky and Brovkin, 1993).

All rates af sapropel after 18 years of application had some effect on the composition of humus substances, but more significant effect was produced by the higher (150, 200 t/ha) rates of sapropel (Table 4). Calcareous sapropel increased the content of humic acids bound with calcium (HA-2) and declined the content of mobile humic acids (HA-1). In this case this regrouping of humic acids fractions was related with the increased amount of calcium incorporated in the soil with calcareous sapropel. The content of total humic acids changed insignificantly. The higher (100-200 t/ha) rates of sapropel reduced the content of fulvic acids significantly. The ratio of humic acids to fulvic acids increased (C_{HA}:C_{FA}=0.74-0.77; in control=0.71). The increased content of insoluble residue was caused by higher rates of sapropel Manure application in every crop rotation improved humus composition, as well (C_{HA}:C_{FA}=0.80) (Bakšiene and Janušiene, 2005).

Table 4. Changes in composition of humus substances under the influence of calareous sapropel and manure

Indices of humus composition	Before experiment	Treatments (after 18 year)						$LS D_{05}$
		NPK (control)	50 t ha^{-1} S	100 t ha^{-1} S	150 t ha^{-1} S	200 t ha^{-1} S	100 t ha^{-1} M	
Organic C (% in soil)	1.13	1.29	1.46	1.52	1.50	1.53	1.52	0.12
C % from organic carbon in the soil								
HA-1	10.6	12.4	8.2	7.9	8.0	7.8	10.5	1.4
HA-2	7.1	6.2	8.9	8.6	8.7	8.5	8.0	1.5
HA-3	10.6	9.3	11.6	11.2	11.3	11.1	11.8	1.6
Sum of humic acids	28.3	27.9	28.7	27.7	28.0	27.4	30.3	1.9
FA-1a	8.0	7.0	5.5	5.3	5.3	5.3	5.9	0.7
FA-1	9.7	7.0	6.2	5.9	6.0	5.9	5.3	1.1
FA-2	7.1	11.6	10.9	10.5	10.7	9.8	11.8	1.6
FA-3	15.0	13.9	16.4	15.8	14.7	14.4	15.1	1.6
Sum of fulvic acids	39.8	39.5	39.0	37.5	36.7	35.4	38.1	1.7
Insoluble residue	31.0	32.5	32.3	34.8	35.3	37.2	31.6	1.8
C_{HA}:C_{FA}	0.71	0.71	0.74	0.74	0.76	0.77	0.80	0.02

S – dry sapropel, M – dry manure

The amount of humus after fertilization with organic fertilizers hardly changed, just like of total nitrogen but its changes were not similar to those of nitrogen. A slightly higher amount of humus was found in treatments where bigger rates of manure had been applied in mixtures. Here humus increased by 0.16, 0.10 and 0.09%. Analysis of soil performed in 2010 revealed that the amount of humus had decreased in all fertilization treatments and was lower than before the start of the investigations. Compared with the background without mineral fertilizers the amount of humus decreased very slightly (0.01-0.08%). Such decrease was noticed in all fertilizing treatments of the investigation. After the second crop rotation, total nitrogen, similarly as humus decreased in all fertilizing treatments. Minimal rates of mineral nitrogen fertilizer were not sufficient to fertilize soil.

9. Application of lakes sapropel for improvement of soil physical properties

In the works of Norwegian and Canadian scientists (Sveistrup et al., 1995, Zebarth et al., 1999) the conclusions about positive effect of lake sediments on physical properties of soil. Introduction of sediments increased the soil porosity and moisture retention capacity of soil, improved the soil texture and quality.

The application of lakes sapropel for soil fertilization was studied in a sandy loam Cambisol (54O491 N, 25O101 E) with a pH value of 6.0, P_2O_5 130-230 and K_2O 150-210 mg kg-1 of soil, humus content 1.7-2.05%. Changes in agrochemical and physical soil properties and the yield of crops fertilized with sediments were studied according to the following experimental design and treatments: 1) control; 2) 10 t ha-1 limestone ($CaCO_3$); 3) 25 t ha-1 calcareous sediments (CS); 4) 10 t ha-1 organic sediments (OS); 5) 40 t ha-1 organic sediments (OS); 6) 25 t ha-1 siliceous sediments (SS); 7) 100 t ha-1 siliceous sediments (SS); 8) 25 t ha-1 calcareous sediments (CS) + 25 t ha-1 manure (M); 9) 10 t ha-1 organic sediments (OS) + 25 t ha-1 manure (M); 10) 25 t ha-1 siliceous sediments (SS) + 25 t ha-1 manure (M); 11) 65 t ha-1 manure (M).

All fertilizers, except for the mineral fertilizers, were applied at the beginning of the rotation, before sowing. During the following years the after effect was observed. All rates of sediments were calculated for dry mass. Minimum rates of mineral fertilizers (N30-60P13-18K42-50) were applied annually before sowing.

To identify changes in physical soil properties soil moisture, bulk density, and total porosity were measured annually after sowing in spring (I) and after harvesting in autumn (II) every year in 1994-2004. Soil moisture, bulk density and porosity were estimated by the weighing method.

9.1. Soil moisture

While analysing the impact of organic lake sapropel upon the soil moisture, it could be observed that the indicators of moisture was lower in spring, and it was lower in autumn (Fig. 20). In this experiment stronger influence was observed when fertilizing with sapropel and sapropel-manure mixture.

To compare with the impact made by the application of manure, the rate of 40 t ha^{-1} of lake sapropel increased the moisture of soil by 1.12–1.57 % units. The effect of manure equalled the smaller rate of organic lake sapropel (10 t ha^{-1}).

The research results revealed that fertilization of sandy loam Cambisol with lake siliceous sapropel exerted a positive effect on the soil moisture (Fig. 20). During all experimental years, it was lower in spring and increased significantly in autumn after harvesting. The effect of the application of sapropel-manure mixture was in many cases equal to that of lower (25 t ha^{-1}) rate of sapropel with mineral fertilizers. The 100 t ha^{-1} sapropel rate (treatment 3), which increased soil moisture by 1.00–1.50 %, was found to be most effective.

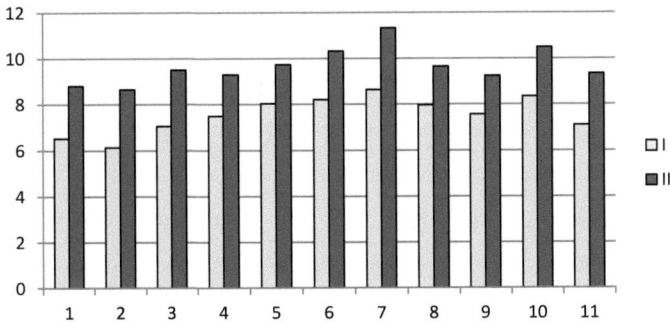

Fig. 20. The effect of lakes sapropel on moisture of sandy loam Cambisol.
(I - spring, after sowing; II - autumn, after harvesting .1-11 treatments of trials: 1) control; 2) 10 t ha⁻¹ limestone (CaCO3); 3) 25 t ha⁻¹ calcareous sapropel (CS); 4) 10 t ha⁻¹ organic sapropel (OS); 5) 40 t ha⁻¹ organic sapropel (OS); 6) 25 t ha⁻¹ siliceous sapropel (SS); 7) 100 t ha⁻¹ siliceous sapropel (SS); 8) 25 t ha⁻¹ calcareous sapropel (CS) + 25 t ha⁻¹ manure; 9) 10 t ha⁻¹ organic sapropel (OS) + 25 t ha⁻¹ manure; 10) 25 t ha⁻¹ siliceous sapropel (SS) + 25 t ha⁻¹ manure; 11) 65 t ha-1 manure (M).

The obtained data shows that fertilization of sandy loam Cambisol with calcareous sapropel, limestone and mixture of sapropel with manure influenced the changes of soil moisture. Even though these rates mostly depend on meteorological conditions and plants that are grown, the data of our research shows that the mentioned features mostly changed because of the amount of organic matter inserted with fertilizers. During the first year of calcareous sapropel application in spring and autumn, the amount of moisture was higher in the treatments where sapropel and mixture of sapropel and manure had been used. It was especially evident in the further years of testing. Each year in spring this difference among treatments increased. Soil moisture increased by 0.67-1.45 % from sapropel application, and mixture of sapropel and manure caused moisture increase by 1.20-2.04 %. This effect lasted for 11 years. Limestone had no impact on soil moisture.

9.2. Soil bulk density

Soil bulk density did not depend upon moisture. The data shows the tendency towards the decrease of this parameter only after fertilization with 40 t ha^{-1} organic lake sapropel rate and sapropel -manure mixture (Fig. 21). Density varied from 1.27 to 1.41 mg m^3 only in certain periods, and just in autumn it was a little bit lower (1.19–1.26 mg m^3). During 11 years the soil bulk density was lower to compare with the data of spring investigations, but sometimes on the contrary, the density was higher in autumn. The soil particle density was not investigated in this experiment because it changes insignificantly under the impact of anthropogenic factors and time.

Analysis of the impact of lake siliceous sapropel on the soil bulk density revealed that sometimes it was changing unevenly. After application of 100 t ha-1 siliceous lake sapropel in the spring the soil bulk density increased up to 1.32–1.35 mg m-3, and the autumn it reduced to 1.06–1.24 mg m-3 .

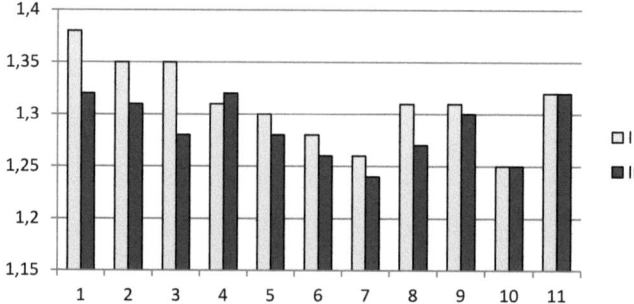

Fig. 21. The effect of lakes sapropel on bulk density of sandy loam Cambisol.
(I - spring, after sowing; II - autumn, after harvesting .1-11 treatments of trials: 1) control; 2) 10 t ha^{-1} limestone (CaCO3); 3) 25 t ha^{-1} calcareous sapropel (CS); 4) 10 t ha^{-1} organic sapropel (OS); 5) 40 t ha^{-1} organic sapropel (OS); 6) 25 t ha^{-1} siliceous sapropel (SS); 7) 100 t ha^{-1} siliceous sapropel (SS); 8) 25 t ha^{-1} calcareous sapropel (CS) + 25 t ha^{-1} manure; 9) 10 t ha^{-1} organic sapropel (OS) + 25 t ha^{-1} manure; 10) 25 t ha^{-1} siliceous sapropel (SS) + 25 t ha^{-1} manure; 11) 65 t ha-1 manure (M).

When soil was fertilized with 25 and 100 t ha-1 rates of lake sapropel, the soil bulk density declined to 1.035–1.20 mg m -3 .

The density of soil depended on the amount of moisture. However, the research data show that density of soil did not notably change and it did not depend on meteorological conditions, moisture of soil and its deepness. Almost in all treatments, this rate was in the range of error (1.30-1.50 mgm3).

9.3. Soil total porosity

Porosity directly depends upon the soil density. With the decrease of density the porosity increases. It is evidently proved by the research results (Fig. 22).

Sometimes total porosity of soil was higher in spring, after soil cultivation, and it decreased in autumn. However, in other years of crop rotation, when perennial grass and fall rye were cultivated, the parameters of total porosity were increasing in autumn. The highest (48.36–57.65 %) total porosity was determined in soil fertilized with 40 t ha^{-1} organic lake sapropel rate almost during the whole period (11 years) of investigation.

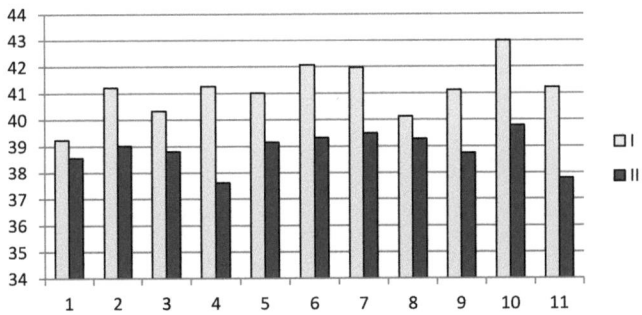

Fig. 22. The effect of lakes sapropel on total porosity of sandy loam Cambisol.
(I - spring, after sowing; II - autumn, after harvesting .1-11 treatments of trials: 1) control; 2) 10 t ha^{-1} limestone (CaCO3); 3) 25 t ha^{-1} calcareous sapropel (CS); 4) 10 t ha^{-1} organic sapropel (OS); 5) 40 t ha^{-1} organic sapropel (OS); 6) 25 t ha^{-1} siliceous sapropel (SS); 7) 100 t ha^{-1} siliceous sapropel (SS); 8) 25 t ha^{-1} calcareous sapropel (CS) + 25 t ha^{-1} manure; 9) 10 t ha^{-1} organic sapropel (OS) + 25 t ha^{-1} manure; 10) 25 t ha^{-1} siliceous sapropel (SS) + 25 t ha^{-1} manure; 11) 65 t ha-1 manure (M).

The total porosity indices, similarly to those of soil bulk density, were changing unevenly due to different fertilizers. In case the bulk density was higher, the total porosity was lower, and vice versa. During the whole experimental period, the total porosity of soil was similar in various levels of arable layer; it was higher only in the autumn and spring. At certain moments the total soil porosity reached 49.01–56.90 and 50.07–55.19 %. A more evident effect was produced by the 100 t ha^{-1} rate of sapropel and the sapropel-manure mixture.

Data of the investigations shows that the amount of organic matter inserted into soil had stronger influence on the change of total porosity. Even though it was established that soil is most porous after the tillage, our researches show that it was when maize and perennial grasses had been grown in the field. In other years, total porosity was higher in autumn after the harvest. After the winter rye and barley were threshed, the total porosity of soil was by 2.45-11.98% higher than that in spring.

Calcareous sapropel is not rich in organic matter, that is why it just slightly influences the total porosity of soil, but in most cases its influence was more evident than of limestones. Nearly in all years, the best rate of total porosity in soil was 48.9-56.2 % after fertilization with calcareous sapropel.

10. Application of lakes sapropel for improvement of soil microbiological properties

Microorganisms are essencial components of soil biocenosis. The species diversity is high, therefore the ecosystem are function normally. Interaction of populations in the microorganism comunity providesthe conditions to utilize various substrates which not be possible in the presence of only one species (Jenzen et al., 1995). Microorganisms of the rhizosphere are particulary important to palnts. They stimulate plant growth by facilitating the assimilation of phosphorus and iron, nitrogen fixation, releasing phytohormones, inhibiting root pathogens, synthesising antibiotics (Glick

Bernard, 1995). The season, soil humidity, pH, fertilization and other factors predetermine the number and species composition of microorganisms in soil.

Sapropel is rich in various microorganisms, especialy in bacteria. Ino ne gramo f dry sapropel the number of protein decomposing can range from several hundred to several millin, the number of bacteria assimilating and of raqnges up to several millions. The number of actinomycetesand cellulose degrading microorganisms is low or they are apsent. Lakes sapropel is much richer inbowls the amount microorganisms increased by 10 times and more (Milto et al., 1981; Soroko, 1985).

Investigations perfomed in Belarus revealed that sapropel induces the activity of microorganismds and simultaneously the biochemical prcesses (Shinkariova et al., 1981).

In Lithuania the effect of sapropel on biological activity of soil is little investigated. Therefore in the Voke Branch of Lithuanian Research Centre for Agriculture and Forestry the research on microorganisms changes in crop rotation on sandy loam Arenosol fertilized with calcareous, organic and siliceous sapropels and its mixtures with other organic fertilizers (manure, sewage) was carried out.

It was revealed that sapropel of different chemical composition predetermined the distribution of certai groups of microorganisms. Sapropel of vasrious chemical composition in most cases stimulated the development of fungi (Fig. 23). Accorging every four years' data the statistical significant positive effect on the amount of fungal propagules had the all kind of sapropel. The diversity of fungal species was greatest in the soil fertilized with organic and especially with siliceous sapropel in comporison with the soil fertilized with calcareous sapropel (Lugauskas et al., 1997; Lugauskas et al, 1999).

The analysis of a species composition of isolated fungi showed that the most numerous were fungal species from genus *Aspergilus, Motierella* and *Penicillium* at first year after application of various chemical composition of sapropel. They dominated in the soil of root zone of crop rotation.

The species of *Penicillium* genus sporulate abundantly and are resistant to unfovorable meteorological conditions. During first four year the number of identical species was low in all treatments of experiment. Only two species *Penicillium* and *Mortierella* were constant components of microorganism communities in the soil fertilized with various types of sapropel.

The species of dominant fungul were increasing every four year. After at the last four year (2006) in the soil were found more number of fungal species: *Acremonium, Mortierella, Penicilium, Clodasporium, Mucor, Fusarium, Verticilium, Sporotichum, Sclerotinia, Alternaria, Trichoderma* (Salina et al., 2001).

Phytopathogenic species of fungi belonging to *Fusarium, Verticilium, Sclerotinia, Altenaria* and other werw spread in certain seasons, but more often their dominance caried temporary and local character and treir population was not high.

The response of fungi of genus *Trichoderma* – typical soil saprophytes – to the addition of fertilizers shows that the influence of sapropel of differnt chemical composition often possesses rather a complex and specific character. During the first year of the experiment the frequency of detection of fungi of this genus were low, buti n subsequent four year of plant growing it decreasing to 20 % in some seasons. Thus about 1/3 of isolated fungi *Trichoderma* were weak vitaly. Obviously during decompossing of various types of sapropel some negatyve effect on the development of separate components of microbial communities is also possible (Salina et al., 2001).

A great number of ammonifying, assimilating mineral N, oligonitrophiles and other physiological groups of bacteria are distributed in sapropel. The total amount of bacteriawas estmated in the tested soil samples. The distinct differences between the distribution of bacteria and plants that were grown in experimental fields were not establishment. The greatest amount of bacteria was detected in the soil fertilized with organic and rather often (especially in dry seasons) – with siliceous sapropel.

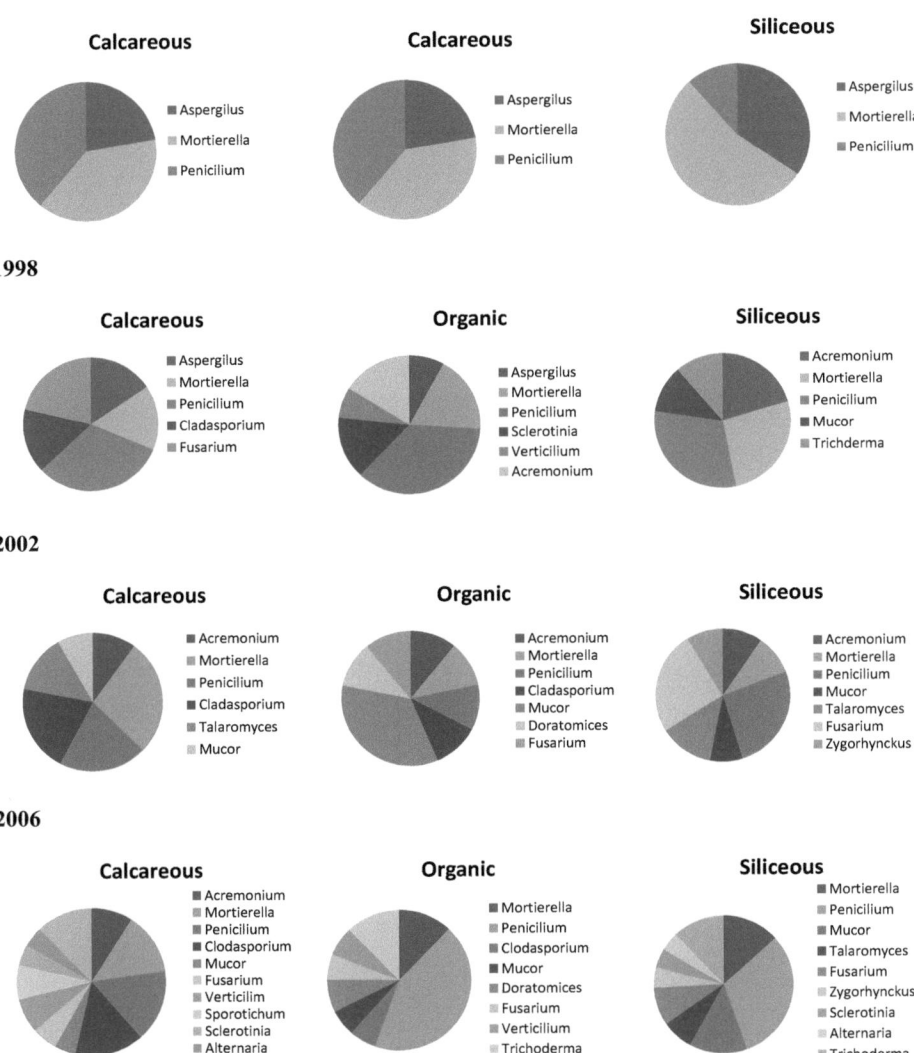

Fig. 23. The effect of lakes sapropel on dominant fungal species of sandy loam Cambisol.

The *Actinomycetes* are very important group of microorganisms in the soil, too. They distinguished themselves by a powerful enzymatic system and took considerable part in the destruction of plant residues. The amont of *Actinomycetes* and other cellulose degrading microorganisms are not very high in sapropel. Nevertheless when sapropel is used for the fertilization it activates the growth of microorganisms and biological activity of soil significantly. Only members of genus *Streptomyces* were counted. At the beginning of the experiment the amount of *Streptomyces* was not great and was similar in all treatments of experimental soil. The greatest amount of streptomyces was detected in the soil fertilized with calcareous sapropel at the end of plat growing season. The efficiency of this type of sapropel on the abundance of streptomycetes in most cases exceeded the efficiency of manure. The positive effect of siliceous sapropel was revealed during dry periods. The greatest total amount of other bacterial genera was estimated in the soil fertilized with organic and siliceous sapropel (Baksiene et al., 2004).

10. Influence of lakes sapropel on the yield

Data on the impact of lake sediments upon the yield of agricultural plants differ. German scientists (Encke et al., 1988; Gruner and Belau, 1990) claim that in all cases of fertilisation with lake sediments the yield of green mass of maize, oats, and fodder cabbages increased by 10-20 %. However, Klimanek A. and Körschens M. (1984) report that lake sediments had a very insignificant impact upon upon the plant yield, for maize the impact was even negative. Russian scientists (Kirdun et. al., 1981; Orlov and Sadovnikova, 1996) in field ezperiment revealed that in sody sandy loam soil lake sediments are not less effecient than peat-manure compost sometimes even superior. The sediment rate of 60-80 t ha^{-1} prodyced the barley yield of 0,34-1,61 t ha^{-1}, while the rate of peat-manure compost produced the yield of only 0,28-1,06 t ha^{-1}. It was also revealed that lake sediments used together with mineral fertilizers augmented the amount of proteins in grain up to 3-5 times, while lake sediments without mineral fertilizers considerably reduced the amount of proteins. Results of the investigations performed in Latvian University of Agriculture revealed that dried, ventilated lake sediments possessing more organic matter can provide the same quantity of yield of potatoes, carrots, and cabbages as manure. Good results were obtained as potatoes, tomatoes, maize were fertilised with frosted sediments. It was established that in lake sediments and in manure the same amounts of nutritious substances available for plants are present (Anspok et al., 1987; Kronbergs and Viduzs, 1993; Liepins, 1993, 1995).

In Voke Branch of Lithuanian Research Centre for Agriculture and Forestry experimental plots for the study of lake sapropel with various chemical composition were established in a field crop rotation (maize, maize (*Zea mays* L.), barely (*Hordeum* L.), with undercrop, perennial grasses (*Trifolium pratense* L. and *Phleum pratense* L.) of the 1st and 2nd year of use, winter rye (*Secale cereale* L.) blend of oats and lupin (*Avena sativa* L. and *Lupinus angustifolius* L.), barley (*Hordeum* L.), with undercrop, perennial grasses (*Trifolium pratense* L. And *Phleum pratense* L.),blend of oats and lupin, barley (*Hordeum* L.).

The application of different rates of various chemical composition of lakes sapropel on a sandy loam soil had a marked wffwct on the increse in the productivity of criop rotation plants.

10.1. Influence of organic sapropel on the yield of crop rotation

The research results show that of maize cultivation on both backgrounds of mineral fertilization sediments rates proportionally enlarged from 10 to 40 t ha^{-1} increased the yield of maize feed units from 2188–2695 to 3028–3255 accordingly (Fig. 24). However, the best and most reliable additional yield (3885 and 4113 feed units) was obtained having applied the mixture of 10 t ha^{-1} sediments plus 25 t ha^{-1} manure. The effect of minimal rates of mineral $N_{60}P_{40}K_{60}$ fertilizers was more evident for maize, in some cases the yield of feed units increased up to 728 (treatment 1).

Productivity of barley grain and straw was mostly effected by the rate of 40 t ha^{-1} lake sediments. On the background without mineral fertilisers reliable extra yield was 2120 feed units and on the background with minimal rates of organic $N_{30}P_{30}K_{50}$ fertilisers – 4021 feed units. Under the impact of sediments-manure and sewage mixture the additional yield differed insignificantly, within the standard error.

Proper fertilization of soil also influenced the yield of grass. Although no statistically reliable additional yield was obtained, the yield of perennial grass of the 1st harvest was rather good even without mineral fertilizers. As the rates of organic lake sediments were increased, the yield of perennial grass (feed units) also increased from 4416 to 5357 (background without mineral fertilizers) and from 4929 to 6284 feed units (background with minimal rates of mineral fertilizers) respectively. Fertilization with 65 t ha^{-1} of manure resulted the same yield as the fertilization with the mixture of 10 t ha^{-1} sediments plus 10 t ha^{-1} manure.

The yield of feed units of perennial grass was larger in the 2nd harvest than in the 1st one, though the additional yields were similar. On both backgrounds of mineral fertilization they increased from 3111 to 3550 and from 3755 to 4063 of feed units and the yield of 3rd grass were bigger and reached 3276-6930 and 4952-7560 feed units depending on the applied rates of sediments. Having applied manure, the yield was obtained the one of similar to control treatments. Whereas the application of sediments-manure and sediments-sewage mixtures produced the same yields as fertilization with pure sediments.

Dry summer of 2004 affected particularly negatively the yield of winter rye cultivated after perennial grass of 1st and 2nd year of use. Grain developed badly and was small. The yield did not reach even 2 t ha^{-1}. If we compare the yield of control treatments of both backgrounds we will see that mineral $N_{40}P_{30}K_{50}$ fertilizers did not help much. However, they were better affected by applied organic fertilizers and, thus, their yield of grain and straw increased for about 100–600 feed units. Yield of blend of oats and lupine on both background depended from the all of applied rates of organic lake sediments and increased from 6128 to 8744 and 6192 to 13248 feed units.

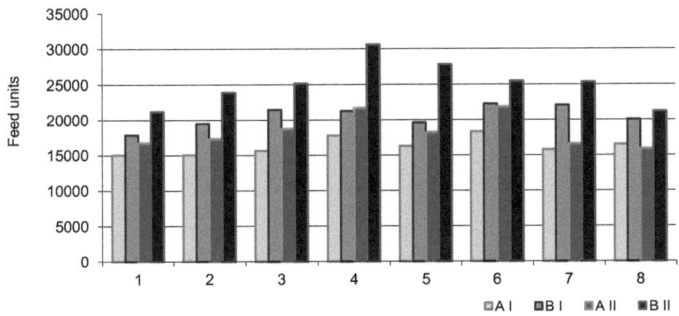

Fig. 24. The effect of organic lake sapropel on yield of crop rotations.
AI - indicators on the background without mineral fertilizers in 1999; AII - indicators on the background of minimum rates of mineral NPK fertilizers in1999; BI - indicators on the background without mineral fertilizers in 2004; BII - indicators on the background of minimum rates of mineral NPK fertilizers in 2004. 1-8 - treatments: 1) control; 2) 10 t ha^{-1} limestone (CaCO3); 3) 25 t ha^{-1} sapropel (S); 4) 1.5 rate of CaCO3 + 25 t ha^{-1} manure (M); 5) 25 t ha^{-1} sapropel (S) +10 t ha^{-1} manure (M); 6) 25 t ha^{-1} sapropel (S) + 25 t ha^{-1} manure (M); 7) 25 t ha^{-1} sapropel (S) + 10 m^{3}sewage (Sw); 8) 65 tha-1 manure (M).

The sum of feed units during the crop rotation of 11 years shows that the rates of organic sediments increased the yield by 3–33 %. However, the mixture of 10 t ha^{-1} sediments plus 25 tha^{-1} manure was most effective on both backgrounds of mineral fertilization. It reliably increased the yield by 22 and 26 %. A rather large and reliable additional yield (24 %) was obtained on the background without mineral fertilizers having applied the largest (40 t ha^{-1}) sediments rate. The effect of manure was the same as the effect of applied lower rates of sediments and sediments-manure mixture.

While analysing the efficiency of the interaction of various sediments rates and its mixtures with other fertilizers it was determined that the yield of feed units in crop rotation depended upon the parameters of soil acidity, when mineral $N_{30-60}P_{30-40}K_{50-60}$ fertilizers were applied for background fertilization and without these fertilizers

10.2. Influence of siliceous sapropel on the yield of crop rotation

The data of the crop yield demonstrates that different rates of lake sediments produced a stronger effect on the productivity of plants during the last year of the crop rotation (winter rye) and in case the sediment rates were 50 and 100 t ha-1 (Fig. 25). In the treatments where the sediment-manure mixture and pure manure were used the yield remained stable. Mineralization of manure had probably taken place earlier, and during the 11th year of these treatments plants were already in a shortage of nutrients. Calculation of the average of feed units per crop rotation revealed that higher rates of sediments (50, 100 t ha-1) augmented the productivity of crop rotation by 8– 30%, while manure by 21–25%. Although the yields were not statistically significant, they were still rather high. Fertilization with lake sediments slightly differs from fertilization with manure. According to Russian scientists, lake sediments are most effective. Their research findings based on sod-podzolic sandy loam soil indicate that sediments were not less effective compared with peat-manure compost, on the contrary, in several cases they were even better.

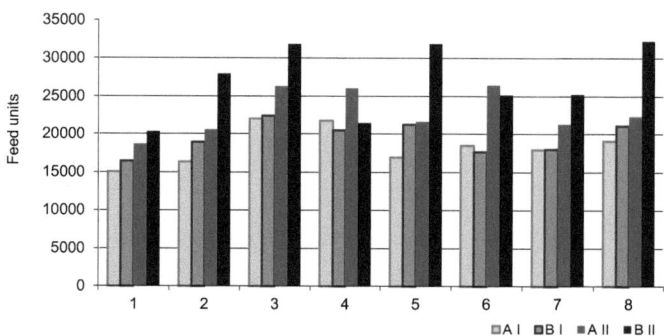

Fig. 25. The effect of siliceous lake sapropel on yield of crop rotations.
AI - indicators on the background without mineral fertilizers in 1999; AII - indicators on the background of minimum rates of mineral NPK fertilizers in1999; BI - indicators on the background without mineral fertilizers in 2004; BII - indicators on the background of minimum rates of mineral NPK fertilizers in 2004.
1-8 - treatments: 1) Control; 2) 25 t ha^{-1} sapropel (S); 3) 50 t ha^{-1} sapropel (S); 4) 100 t ha^{-1} sapropel (S); 5) 25 t ha^{-1} sapropel (S) + 10 t ha^{-1} manure (M); 6) 25 t ha^{-1} sapropel (S) + 25 t ha^{-1} manure (M); 7) 25 t ha^{-1} sapropel (S) + 10 m^{3}ha^{-1} sewage (Sw); 8) 65 t ha^{-1} manure (M).

10.3. Influence of calcareous sapropel on the yield of crop rotation

The application of different rates of lake sapropels on a sandy loam soil had a marked effect on the increase in the productivity of crop rotation (Fig. 26). Comparison of the data of different treatments, where soil had been only limed and fertilized with calcareous sapropel alone (3 treatment), revealed that in most cases fertilization with sapropel during the crop rotation in both backgrounds was more effective, it produced by 2402 and 3451 more feed units than liming. Analogical results were also in treatments where mixtures of limestone and manure had been used for fertilization (4 treatments) or the mixture of sapropel and manure (5, 6 treatments). Mixtures of sapropel and manure as well as of limestone and manure had the strongest influence on the productivity of crop rotations. In both mineral fertilizer backgrounds 25 t ha-1 sapropel + 10 t ha-1 manure gave statistically reliable supplements of feed unit yield (4154 and 3727 feed units). In most cases, bigger feed unit supplements (2402 and 3451 feed units) were obtained in the treatments where less manure

had been used to mix with sapropel (10 t ha-1), but not bigger (25 t ha-1). During the crop rotation, mixture of 25 t ha-1 dry sapropel and 10 t ha-1 of manure, produced a reliable feed unit supplements (4154 and 3727). It seems that even a low amount of manure is adequate to activate the process of sapropel mineralization. Yet, 10 m3 of sewage was not enough. Efficiency of sapropel and sewage mixture (7 treatments) was similar to the effect of sapropel alone.

Analysis of the data on productivity of crop rotation after application of calcareous sediment and limestone affirmed that in most cases, fertilization with calcareous sapropel was more effective than application of the limestone. On the basis of this data, it is possible to state that the calcareous sediment can serve not only as calcareous matter, but it is also a source of nutrients for plants.

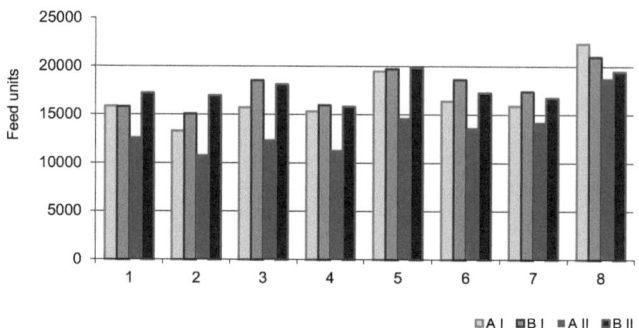

$\square$ A I $\blacksquare$ B I $\blacksquare$ A II $\blacksquare$ B II

Fig. 26. The effect of calcareous lake sapropel and limestones on yield of crop rotations.

AI - indicators on the background without mineral fertilizers in 1999; AII - indicators on the background of minimum rates of mineral NPK fertilizers in1999; BI - indicators on the background without mineral fertilizers in 2004; BII - indicators on the background of minimum rates of mineral NPK fertilizers in 2004. 1-8 - treatments: 1) control; 2) 10 t ha^{-1} limestone (CaCO3); 3) 25 t ha^{-1} sapropel (S); 4) 1.5 rate of CaCO3 + 25 tha^{-1} manure (M); 5) 25 t ha^{-1} sapropel (S) +10 t ha^{-1} manure (M); 6) 25 t ha^{-1} sapropel (S) + 25 t ha^{-1} manure (M); 7) 25 t ha^{-1} sapropel (S) + 10 m^{3}sewage (Sw); 8) 65 tha-1 manure (M).

The introduction of different rates dry sapropel in sandy loamy soil has corresponding influence on the increase of productivity of plants in four crop rotations.

The cumulative curve of the increase of plant productivity (Fig 27) showed that during first four years of crop rotation the efficiency of the smallest rate (50 tha^{-1}) of sapropel was equal to the efficiency of manure.

Higher rates (100-200 tha^{-1}) of sapropel proportionally increased the productivity of crop rotation. However, during the years when potatoes and oats were cultivated in the crop rotation, all rates of sapropel reduced the productivity of these crops.

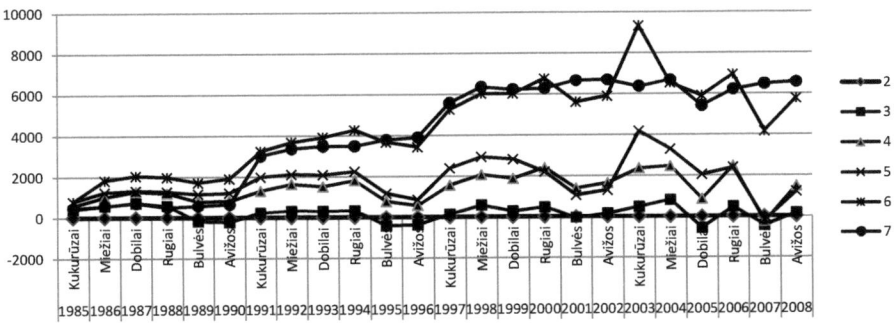

Fig. 27. Cumulative curve of yield increase from the different rates of calcerous sapropel.

2-7 - treatments of trials: 2) NPK (B-background, 3) B + 50 t ha^{-1} sapropel (S), 4) B + 100 t ha^{-1} sapropel (S), 5) B + 150 t ha^{-1} sapropel (S), 6) B + 200 t ha^{-1} sapropel (S), 7) 100 t ha^{-1} manure (M).

In the second crop rotation the productivity increased from the repeated introduction of manure and further effect of the largest rate (200 tha^{-1}) of sapropel was even more efficient. However, at repeated cultivation of potatoes and oats, the negative effect of sapropel (50 tha^{-1}) on the productivity reappeared, like in the first crop rotation. The data show that from the beginning of the second (1991), third (1997) and fourth (2003) crop rotation the cumulative curve began to rise but in 1995 - 1996, 2001 - 2002 and 2007-2008 when were cultivated potato and oats, the productivity, like in previous crop rotations, decreased. These crops are more tolerant to acid soils. Large amounts of Ca++ ions were introduced into the soil with the high rates of calcareous sapropel. They might have blocked up available potassium and other elements necessary for the development of potato and oats yield.

There are established that organic and siliceous sapropel influences agricultural crops much more effectively (Orlov and Sadovnikova, 1996). However, the data of the experiments provided in this article show, that in the majority of cases calcareous sapropel gives an increase in crop production, too.

References

Andresini, A.; Loiacono, F.; De Marco, A.; Spangoli, F. Recent Sedimentation and Present Enviromental State of "Lesina lake". Proceedings of the International Conference on Southern European coastal lagoons: The Influence of River Basin-Coastal Zone Interactions. Perrara, Italy: 2003, 51.

Anspok P.I., Skromainis A.A., Grinberg V.P., Dubrovskaja D.A. Sapropel – a perspective form of organic fertilizer. Chemistry in agriculture. 1987. 4: 30-33. (in Russian).

Bakšienė E., Repečkienė J., Salina O. Succession of agrochemical properties and microorganisms in sandy loam Cambisols fertilized with sapropel of different chemical composition. Water management Engineering, 2004. 27 (47): 37–41. (in Lithuanian).

Bakšiene, E., Janušiene, V. The effects of calcareous sapropel applicationon the changes of Haplic Luvisols chemical propertiesand crop yield. Plant, Soil and Environment. Dev. 2005. 51(12): 539–544.

Bakšienė E., Fullen M. A., Booth C. A. Agricultural soil properties and crop productionon Lithuanian sando and loamy Cambisol safter the applicatio no fcalcareous sapropel fertilizer. Archives of Agronomy and Soil Science, 2006. 52 (2): 207–222

Bakšienė E. The effects of lakęs sapropels on the chantes in sandyloamc ambisol properties. Environmental Engineering, 2008. 1: 29–34

Bakšienė E., Ciūnys A. Dredging of lake and application sapropel for improvement of light soil properties. Journal of Environmental Engineering and Landscape Management, 2012. 20, Iss. 2: 97–103.

Baksienė E., Asakaviciute R. The effects of organic lake sediments on the crop rotation yield and soil characteristics in Southeast Lithuania. Applied Ecology and Environmental Research, 2013. 11 (4): 557–567.

Belenkij S.G. Genesis and age of sapropelic deposits of a peat field Natural boundary Kamensky Bridge. Problems of use of sapropels in a national economy. 1981: 11-12. (in Russian).

Bieliukas K. Lakes of Lithuanian SSR. Vilnius, 1956, 51 p. (in Lithuanian)

Bieliukas K. Fundamentals of lakes researches. Vilnius: Mokslas, 1961, 358 p. (in Lithuanian).

Booth, C. A., Baksiene, E., Fullen, M. A., Ciunys, A. Long-term agrochemical dynamics: engineering, application and challenges of calcareous sapropel as a soil fertilizer. International Journal of Ecodynamics. Dev. 2007. 2: 108-116.

Chochlova, O. B. Sapropel - Ameliorants and Fertilizer of Long Action. The author's thesis of Agricultural Sciences. Moscow, 1997. 20 p. (in Russian).

Cherkinsky, A.E.; Brovkin V.A. Dynamics of radiocarbon in soil. Radiocarbon, 1993. 35 (3): 363-367.

Ciūnys A., Lazauskienė L., Katkevičius L. Sapropel – our treasure. Vilnius: Baltic ECO,1994. 29 p. (in Lithuanian)

Ciūnys A., Katkevičius l. 2008. Environmental work and their regulation. Kaunas: Ardiva, 57 p.

Dudin V.M., Jermakov M., V., Klimovickij M., L. Use of sapropel of the lake of Nero in feeding of animals. Improvement of natural lakes and artificial water reservoirs. Kaunas-Akademija, 1997, 103-104. (in Russian).

Encke O., Korschens M. Dungung Seeschlamm auf humusverarmten Sandbode // Einfluß auf bodenphysikalische Eigenschaftenund Ertag. Arch. Gertenbau. 1988. 36(8): 501-508.

Garunkštis A. Sedimentation processes in the lakes of Lithuania. Mokslas, Vilnius, 1975, 295 p. (in Russian)

Garunkštis A., Stanaitis A. Why is drain the lakes of Lithuania. Vilnius:Mokslas, 1978. 92 p. (in Lithuanian).

Garunkštis A. Lithuanian water. Mokslas, Vilnius, 1988. 192 p. (in Lithuanian)

Gediminas A., Kanopkaite-Rozgiene S., Borodina T. Comparative effect of sapropel and biomitsin on growth and some indicators of pigs. Works of Sverdlovsk agricultural institute, 1962, 10: 361-362.

Glick Bernard R. The enhancement of plant growth by free-living bacteria. Canadian Journal of Microbiology. 1995. 41(2):109-117.

Grantina-Ievina L., Karlsons A., Andersone-Ozola U., Ievinsh G. Effect of freshwater sapropel on plants in respect to its growthaffecting activity and cultivable microorganism content. Zemdirbyste-Agriculture, 2014, 101(4): 355–366.

Grigorov M.S., Ovchinnikov A.S. Utilization of Volga-Akhtuba flooplain sapropel. Pochvovedenie, 1994, 0(5), p. 62-66 (in Russian).

Grishina L.A., Kurmisheva N.A., Kazakova S. V., Moroz O.R. The effect of overwashing of sapropel fertilizers on agrochemical propeties and humus status of soddy-gley soil. Moscow University soil science bulletin. 1990, 45(2): 62-69. (in Russian).

Gurskis V., Navickas J. Application of sapropel for made materials of manufacture. Management of rural development according of EU's example. LUA scientific conference. 2001, 31-32.

Gruner A., Belau L. Zur Dauerwirkung von Schlamen aus Seen und Brackgewassern auf den C- und N- Gehalt des Bodens. Tagungsber. 1990. 295: 207-214.

James W.F., Eakin H.L., Barko J.W. Manipulation of sediment nitrogen via dewartering and rehydration: Implications for macrophyte control and nitrogen disipation. ERDC/TN APCRP-EA-06:12.

Jenzen R.A., Cook F.D., Mc Gill W.B. Compost extract added to microcosms may stimulate community level controls on soil microoganisms i nelement cycling. Soil Biology and Biochemictry. 1995. 27(2): 181-188.

Kabailienė M. Structure and evolution of lakes and wetlands. Stone Age in Southern Lithuania. 2001, 121-125. (in Lithuanian)

Kabailienė M. That tell the lake sediment deposits. Science and life. 2006. 11: .22-23. (in Lithuanian)

Kavaliauskienė J. Present trophic condition of the Lithuanian lakes and quality of water on phytoplankton indicators. Improvement of natural lakes and artificial water reservoirs. Kaunas-Akademija, 1997, 8-10. (in Lithuanian).

Kanopkaite-Rozgiene S., Pakarskyte K., Gediminas A. About stability of vitamin B_{12} in sapropel. Works of Sverdlovsk agricultural institute, 1962, 10: 201-208. (in Russian)

Kasperiūnaitė D., Mikuckis F., Navickas J. Investigation of Thermophysical Properties of Unburnt Clay Samples Having Sapropel Additives. Science of Technology. 2010. 86 (39): 71-77. (in Lithuanian).

Katkevicius, L., Ciunys, A. and Baksienė, E. The Sapropel of Lakes for Agriculture. LZI, 1998. 94 p. (in Lithuanian)

Kilkus K. The lakes of Lithuania and their use in a national economy. Vilnius: Žinija, 1987, 29 p. (in Lithuanian).

Kilkus K. Researches of lake hydrology at Vilnius University. 7 World Lithuanian Symposium on Sciences and Arts Geography section reports. Vilnius, 1991, p. 31-36. (in Lithuanian).

Kilkus K. General hydrology (lakes and water reservoirs).Vilnius: Arėjas, 1993. 96 p. (in Lithuanian).

Kilkus K. The geography of the Lithuanian waters. Vilnius University, Institute of Geography. 1998, Vilnius: Apyaušris. – 248 p. (in Lithuanian).

Kireycheva, L. V. and Chokhlova, O. B. Comparative efficiency of humic "Darina" preparations and sapropel. Agrochemical Messenger. 2004. 3:19-20. (in Russian).

Kirdun E.A., 1981. Efficiency of sapropelic fertilizers on podsolic soils. Problems of use of sapropels in a national economy. 1981: 112-114 (in Russian).

Klimanek E.-M., Körschens M. Mineralisierungsverhalten und Stickstoffereitstellung. Zbl. Mikrobiology. 1984. 561-568.

Kronbergs A., Viduzs A., 1993. The value of organic fertilizers and sapropel. Raziba. 1993. 8: 24-27.

LAND 20-2005. Requirements for Sewage Sludge Application in Fertilization on recultivation. Vilnius. LRAM, 2005. Pp. 8.

Liepins J. The use of sapropel for soil ameliaoration. Latvijas Lauksaimniecibas universitetas raksti. 1993. 277(1): 72-74.

Liepins, J. Advantage of sapropel for the improvement of sandy soils with a low humus content. Lauksaimniecibas universitetas raksti, 1995. 278(1):8-9.

Linčius A. Sapropel of Lithuania and his prospect. Geographical chronicle. 1977, 15: 103-112. (in Lithuanian).

Liuzinas R., Jankevicius K. Refreshing lakes. Science and life. 2005, 11: 29-32. (in Lithuanian).

Lopotko, M. Z., Evdokimova, G. F. Sapropels and products on their basis. Minsk, 1986: 112-146 (in Russian).

Lopotko, M. Z., Evdokimova, G. F., Kuzmitsky, P. L. Sapropel in Agriculture. Minsk, 1992, 140 p. (in Russian).

Lugauskas A., Salina O., Bakšienė E. The changes of mikromicetes in the soil fertilizing with sapropel. Improvement of natural lakes and artificial water reservoirs. Kaunas-Akademija, 1997, 95-98. (in Lithuanian).

Lugauskas A., Repečkiene J., salina O., Bakšienė E. Microorganisms in soil of clover-field with sapropel. Botanica Lithuanica, 1999, 5(2): 171-181.

Malaškaite B. Efficiency of feeding of sapropel and microelemnts at sagination of pigs. Works of Sverdlovsk agricultural institute, 1962, 10: 319-328. (in Russian).

Mazvila, J., Vaicys, M., Buivydaite, V. Macromorphogical Diagnostic of Lithuania's Soils. LZI, 2006: 284 p. (in Lithuanian)

Milto N.I., Vorochaeva I.E., Karbanovich A.I. Microflora of sapropels and sapropel - manure composts. Problems of use of sapropels in a national economy. 1981: 51-52. (in Russian).

Mikulionienė S., Triukas K., Valius M. Sapropel of feed for cattle. Agriculture. 1998. 8 (9): 26. (in Lithuanian)

Mikulionienė S., Baležentienė L. Chemical composition and influence of sapropel on live weight gains in fattening pigs. Veterinary and Zootechnics (Vet Med Zoot). 2009, 48 (70): 37-44. (in Lithuanian)

Mud therapy in Russia sapropels. medicalency.com/grjazelechenie.htm (14/07/2015).

Organic Fertilizers Based on Sapropel and Peat. (http://www.latpower.lv/index.php?page=2) (14/07/2015).

Navickas J., Gurskis V. The influence of organic and calcareous sapropel on samples properties of not calcinated clay. Vagos. 2005, 67(20): 95–100. (in Lithuanian).

Orlov, D. S. and Sadovnikova, L. K. Nontraditional ameliorants and organic fertilisers. Euroasian Soil Science, 1996. 29(4): 474-479.

Ostrovskij, M. V. Testing HUMIN PLUS Microfertilizer. European Agrophysical Journal, 2014. 1(2), 77 – 83.

Ostrovskij, M., Krueger E., Kurtener D., Tsukanov, S. V Maklyuk E. Application of Organic Fertilizers Based on Sapropel and Peat in Countries of Middle East. European Agrophysical Journal, 2014, 1(3): 118 – 128.

Ozeraitienė, D., Plesevicienė, A. K. and Gipiskis, V. Duration of action of different lime fertilizers and their effect on soil properties and yield of crop rotation crops. Zemdirbyste-Agriculture, 2006. 93(1):3-21 (in Lithuanian).

Paliulis D. Assesment of lake bottom sediment pollution by lead and cadmium. Polish Journal of Environmental Study. 2014, 23 (4): 1273-1279.

Roberts N., Eastwood W.J., Kuzucuoglu C., Florentino G., Caracuta V. Climatic, vegetation and cultural change in the eastern Mediterranean during the mid-Holocene environmental transition. The Holocene, 2011. 21: 147-162.

Rubinstein, A. J. Typology and Classification of Fresh-water Lake Carbonate Deposits. Science and Production Integration in Branches of Agro-industrial Complex, Vilnius, 1984. 100 p. (in Russian).

Salina O., Repeckiene J., Lugauskas A., Baksiene E. Variation of microorganisms in Haplic Arenosol fertilized with sapropel of various chemicak composition. Agriculture. Scientific articles. 2001. 76: 87-102.

Samutin. N. Experience in using sapropel mud in combination with a magnetic field in treating cervical osteochondrosis. Vopr Kurortol Fizioter Lech Fiz Kult. 1997 Sep-Oct;(5):25-6. (in Russian)

Sapro Elixir. http://www.huminvit.lv/sapro-elixir?lang=eng (12/06/2015).

Sapropeat   Uni. http://athdevelopments.co.uk/downloads/Humin%20Plus%20Brochure.pdf (12/06/2015).

Sapropels of BSSR. Classification industrial and genetic, introduction of 01.01.1986. Minsk, 1986, 3 p. (in Russian).

Shinkariova T.A., Kurbatova-Blikova N.M., Raxubo T.A. Microflora of some Belarusian sapropels. Problems of use of sapropels in a national economy. 1981: 65-67. (in Russian).

Soroko V.I. Change of biological activity of soddy podsolic sandy loam soil at use of sapropelic fertilizers. Soil science and agrochemistry. 1985: 71-75. (in Russian)

Stankevica, K., Klavins, M., Rutina, L. Accumulation of Metals in Sapropel. Materials Sciences and Applied Chemistry, 2012. 26:.99-105.

Stepanova, E. A., Orlov, D. S. Chemical Characterization of Humic Acids from Sapropels. Eurasian Soil Science. 1996. 29(10): 1107-1112.

Sveistrup, T., Marcelino, V. and Stopps, G. Effects of slurry application on the microstructure of the surface layers of soils from northern Norway. Norwegian Journal of Agricultural Sciences, 1995. 1-2: 1-13.

Taminskas J. Features of the Karst regional water management. Northern region karstic Lithuania (geographical aspects of environmental management), 2000: 137-161. (in Lithuania)

Tarcijonaitė M., Miknevičius M. Lakes sludge – and cosmetics. News of business, 2014: 10.

Valius M., Kukenis D., Linkus S. Experience of feeding of sapropel to pigs and cockerels. Works of Sverdlovsk agricultural institute, 1962, 10: 361-362. (in Russian).

Zaccone C., Cocozza C., Shotyk W., Miano T. M. Humic acids role in Br accumulation along two ombrotrophic peat bog profiles. Geoderma. 2008. Vol. 146. P. 26–31.

Zebarth, B. J., Neilsen, G. H., Hogue, E. and Neilsen, D. Influence of organic waste amendments on selected soil physical and chemical properties. Canadian Journal of Soil. 1999. **79**(3):399-409.

Žvironaitė J.; Ciūnys A.; Gerdžiūnas P. Research of opportunities for application of lakes cleaning product – sapropel. Environmental Engineering.2002, X (4): 168–175. (in Lithuanian).

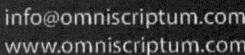

Printed by Books on Demand GmbH, Norderstedt / Germany